National Environmental Policy Act (NEPA) Process

Author

Keshava S. Murthy
Research Leader
Department of Earth Sciences
Battelle Northwest Laboratories
Richland, Washington

CRC Press, Inc.
Boca Raton, Florida

Library of Congress Cataloging-in-Publication Data

Murthy, Keshava S.
National Environmental Policy Act (NEPA) process/author Keshava S. Murthy
p. cm.
Bibliography: p.
Includes index.
ISBN 0-8493-6746-8
1. Environmental law — United States. 2. Environmental impact statements — Law and legislation — United States. I. Title.
KF3775.M83 1988
344.73'046 — dc19
[347.30446]

87-22197
CIP

This book represents information obtained from authentic and highly regarded sources. Reprinted material is quoted with permission, and sources are indicated. A wide variety of references are listed. Every reasonable effort has been made to give reliable data and information, but the author and the publisher cannot assume responsibility for the validity of all materials or for the consequences of their use.

Direct all inquiries to CRC Press, Inc., 2000 Corporate Blvd., N.W., Boca Raton, Florida, 33431.

International Standard Book Number 0-8493-6746-8

Library of Congress Card Number 87-22197
Printed in the United States

FOREWORD*

Since the advent of printing, it has been the prerogative and frailty of authors to feel sanguine that their technical and scientific knowledge and know-how, however meager, could be of interest and use to others in the field. Hoping that I am only partially guilty of this frailty rather than a prerogative, I offer this book to you, the reader. I ask that you be the judge of its value. Also, I sincerely welcome your comments about the book.

This book is mainly about the National Environmental Policy Act (NEPA) of 1969 and the process for its implementation. This is most often called the ''NEPA process''. The U.S. Congress legislated NEPA in response to the public awareness of the damage to the environment caused by the immense industrial activities of man, heedless of the resulting environmental pollution, quite often irreversible. NEPA requires decision makers in government and industry to harmonize all aspects of a project or activity: technological, environmental, economic, sociopolitical, and the continued survival of the ecosystem. In short, NEPA advocates that decision makers approach industrial development holistically by balancing the environmental, economic, and technological factors of projects to protect and enhance public health and welfare. NEPA should be viewed as an essential survival tool, not as a mere law that must be obeyed in letter. Nevertheless, the makers of this law and its practitioners could feel like Prometheus, who stole fire from heaven, gave it to man, but was tortured for the favor.

Like NEPA, the tone and spirit of this book are not meant to be merely environmental and technical, but also ethical, altruistic, political, and social. In short, the overall approach is holistic and asks the readers to use a comprehensive fervor in reading and applying its contents to decision making.

I believe that the book reflects the accumulated experience of my work, which I began in 1969, in the environmental protection business. I started out as an air pollution chemist in the city of Alexandria, Virginia. The pollution control field attracted me because I felt that, as children of nature, men must protect nature even in using up her stores for our ever increasing need for comfort, i.e., exploit nature, but do so with an enlightened self-interest! To work in the environmental protection field was, to me, both a spiritual experience and a method of contributing to the continued survival and quality of life of the inhabitants of this planet and their progeny.

I have tried to incorporate in this book the 17 years of experience gathered from my work in various capacities and fields: (1) as environmental engineer for Montgomery County, Maryland, (2) as a U.S. Environmental Protection Agency (U.S. EPA) scholar at the University of Cincinnati, (3) as senior research scientist at the Battelle Columbus Laboratories in Columbus, Ohio, and (4) as staff scientist and project manager at the Battelle Pacific Northwest Laboratories in Richland, Washington. Specific experiences have been distilled into this book from my work in the following areas: (1) environmental assessment, systems analysis, and program support for the fluidized-bed combustion process in support of studies for the U.S. EPA; (2) evaluation of flue gas desulfurization processes; (3) evaluation of reductant gases for flue gas desulfurization; and (4) environmental compliance documentation for a large, coal-fired power plant and for nuclear fuel processing operations. Also, the hands-on laboratory work and hardware experiences at Battelle in the U.S. and the National Chemical Laboratory and the Indian Institute of Petroleum in India have provided me the necessary confidence to write this book.

Perhaps the most significant factor urging the creation of this book is the very positive feedback from my students in the short course ''Environmental Compliance Documentation

*** The contents of this book and the views expressed herein are the sole responsibility of the author (Keshava S. Murthy) and not of any other individual, institution, or organization, including Battelle Northwest Laboratory of the Battelle Memorial Institute.**

in Project Planning and Management'' offered at the Tri-Cities University Center, Richland, Washington. The Center is operated under the joint auspices of the University of Washington, Oregon State University, and the Washington State University.

The need for this book arises because, although the awareness for faithful implementation of NEPA has now matured, the method of NEPA implementation is not always known and, hence, needs to be clearly laid out for the benefit of project engineers and the decision makers in government and industry. This book aims to do that. The methods presented in this book have established a demonstrated success in many projects which are now operational. I have borrowed freely from publicly available documents generated by the projects that I directed. I have also borrowed from my experiences, that are a part of my know-how and are not recorded anywhere. These experiences have accumulated in my mind not only from project work, but more importantly during the give-and-take in the classroom at the Tri-Cities University Center. The students identified what areas they wanted covered in the course curriculum and these areas are now included in this book.

The process of implementing NEPA is a multidisciplinary exercise that requires cooperation and coordination among numerous disciplines, personalities, and a blending of their knowledge to produce an analysis that would support an optimum decision. I have tried to describe in this book how this multidisciplinary effort can be achieved.

The subject and contents of this book (which I have summarized later in the introduction chapter) may be useful as a teaching tool in the undergraduate and postgraduate curricula. The subject can be taught in a 3 or 4 credit hour course in a quarter. Such a course may, perhaps, be called ''Environmental Compliance Aspects of Decision Making'' or ''The NEPA Process in Project Design and Implementation''. I feel that this subject will be helpful to students in Business Administration (MBA), Systems Management, Construction Management, Industrial Engineering, Chemical Engineering, and Nuclear Engineering at the senior level or postgraduate level. However, the book as now designed is suited more to a curriculum that teaches **overall methodology with some course exercises scattered here and there**, and less suitable for an in-depth study of estimating environmental impacts, modeling calculations, or control equipment design, each of which is already a subject of in-depth study in colleges. The book integrates the purpose of all these latter courses and give a philosophical answer to the ''why'' of these other studies that students most often seek.

Will NEPA usher in an environmental Utopia on Earth? Probably not completely; but if NEPA is ignored, it is certain that environmental disaster will not be far away. Therefore, quite probably, if all of our potential decision makers of the future study NEPA implementation in their early educational phase, America and the world will be spared what Emerson's Webster would perhaps have said today:

Why did all Nature's gift to man fail? He wounded Earth, heedless of her forbearing wail!

NEPA may yet prevent the import of this sad song.

THE AUTHOR

Keshava (Kesh) S. Murthy received his masters degree in Environmental Engineering from the University of Cincinnati in 1974 and his B.S. in Chemical Engineering from the Indian Institute of Chemical Engineers in 1966. In 1958, he received a B.S. in Chemistry from the University of Mysore, India. Currently, Mr. Murthy is a senior project manager and staff scientist at the Battelle Memorial Institute's Northwest Laboratory, where he directs projects involving the preparation and review of environmental compliance documentation of projects in nuclear fuel processing, commercial nuclear waste disposal, defense high-level nuclear waste storage tank construction and operations, and coal combustion. He teaches a 2-day short course on environmental compliance documentation at the Tri-Cities University Center under the joint auspices of the University of Washington, the Oregon State University, and the Washington State University.

Mr. Murthy worked on various research projects at the Battelle Memorial Institute's Columbus Laboratories during 1974 to 1978. Among these: a laboratory-scale experimental project to discover a solid solvent for high-temperature desulfurization of coal-gas; support to EPA on environmental assessment, systems analysis, and program support for fluidized-bed combustion; flue-gas desulfurization process assessments; cost of clean air; identification of emerging technologies; impacts of oil shale development; and reductant gases for flue gas desulfurization. He also has 7 years of experimental work experience in ethyl acetoacetate process development and petroleum refining process development. Mr. Murthy has published research articles in *Environmental Science and Technology, CHEMTECH, Journal of Air Pollution Control Association, Chemical Engineering Progress,* and other scientific journals. He is an active member of the American Nuclear Society, including organizing and chairing technical sessions for national meetings, and technical chairman of environmental sciences division of the ANS. Mr. Murthy is married, has a son and a daughter, and emigrated to the U.S. in 1968.

ACKNOWLEDGMENTS

I sincerely thank the many colleagues at the Battelle Memorial Institute who offered help, encouragement, and constructive suggestions over the past 8 years in my efforts to develop a successful approach to implementing the National Environment Policy Act. I can call these efforts successful because the projects for which the NEPA process was implemented have all been successfully operating as planned. One such project, the construction and operation of a large, coal-fired power plant, has been used as an example in this book. I have borrowed freely from this example from the published environmental impact statement of the Federal Energy Regulatory Commission (FERC). I wish to thank the FERC and the many other sources of published literature including the Council on Environmental Quality, the *Federal Register,* and the Department of Energy. The Tri-Cities University Center in Richland, Washington, has helped me to teach "Environmental Compliance Documentation in Project Planning and Management", receive feedback from students, and learn therefrom. More than 100 students provided positive and constructive suggestions to improve that course material, which is incorporated into this book. I wish to extend my sincere thanks to both the Center and the students for their help.

I am thankful for the privilege of participation in and direction of many environmental protection projects staffed by creative, dedicated, and untiring scientists, engineers, editors, word processors, and other supporting workers. The exemplary quality of these persons and the creative atmosphere at Battelle is a tribute to Gordon Battelle's vision and to the concrete manifestation and the able management of his dream, the Battelle Memorial Institute.

I am particularly thankful to Dr. John B. Brown, Jr. (Program Manager for Environmental and Risk Assessment at Battelle) and to Mr. A. M. Sastry (quality control and quality assurance expert and systems management specialist) for their specific suggestions to improve the book. This book has had the benefit of a critical review by Ms. Dinah Bear, General Counsel, Office of Environmental Quality in the Executive Office of the President, Washington, D.C. Also, Dr. Fred L. Rose, Director of Continuing Education, Tri-Cities University Center, provided useful reviews. I am deeply grateful to both Ms. Bear and Dr. Rose for their interest, time, and effort. I own the good and the bad in this book. I hope that the readers would send their suggestions for the improvement of this book to me, in care of CRC Press.

To
Prema, love's exemplar;
Ben Ridgely, who made the impossible, possible;
Judge Bowen, for his faith and prop.

TABLE OF CONTENTS

Chapter 1

INTRODUCTION, ORGANIZATION, AND SUMMARY

I. INTRODUCTION

The *National Environmental Policy Act Process* (NEPA process) offers a comprehensive and mature method of decision making to the leaders in government and industry. The NEPA process has matured into a useful decision-making technique during the past 17 years. It involves decision making after deliberate evaluation of the technical-economic-environmental-political aspects (with emphasis on environmental impacts) of any major action that would commit the use of the resources of the earth in a major way. In fact, NEPA began because of the question posed by many concerned Americans in the 1960s: is man progressing or regressing? The question may appear philosophical, but it is actually a realistic worry we should face squarely. The human population is constantly increasing, from about 340 million in 1000 A.D. to about 5 billion in 1986 A.D. Human wants and needs have also similarly multiplied. Human life expectancy has increased. These factors have forced the human being to use the resources of the earth at a much faster rate than ever before in our known history.

While human population has visibly increased, the wildlife and plant populations have decreased alarmingly in the past 100 years. Several animal species, including such magnificent creatures as the eastern wapiti, Audubon's bighorn, the graceful passenger pigeon, and the Carolina parakeet, have become extinct, and several more are diminishing to the point of extinction.[1] The habitat of the biological population is being taken over by the human population. About the year 1900, North America was still an Eden swarming with stags, Virginian deer, bison, and bears. In 1986, most of these species are nearly gone. More than 205 species, including mammals, birds, reptiles, amphibians, and plants, are in danger of becoming extinct throughout its natural habitat.[6]

The level of carbon dioxide in the air has increased steadily from 315 ppm in 1958 to 343 ppm in 1986, increasing the threat of the greenhouse effect that could either warm the earth, causing widespread changes in climate, or cool the earth and drastically reduce agricultural output while increasing the already-growing energy consumption. The world population during this period increased by about 2 billion. World-wide releases of chlorofluorocarbons (like freon, used as a refrigerant) to the atmosphere has increased from zero in 1930 to 12 billion lb, contributing to the fear of depletion of the protective ozone layer. The use of pesticides has increased from 286 million lb in 1950 to about 1 billion lb in 1983, increasing the threat of groundwater pollution. These are some of the negative environmental events that happened without the direct knowledge of most of the human inhabitants of this planet; their consequences are still in the formative stages.

Contrary to these scientific facts, what most of us experience today in everyday America is quite positive and pleasant. We get up in the morning and turn a handle and hot water starts flowing freely to soothe us with hot showers; no more woodchopping to prepare the water. Milk is always ready in the refrigerator for our breakfast. No matter what the season, orange juice, fresh fruits, fresh vegetables, and plenty of other foods are readily available and close at hand. The weather outside is of no consequence because the weather in the home and in the car is always comfortable. So you, the reader, may wonder why even ask the rhetorical question, "Is man progressing or regressing?"

As thinking people we always must wonder, "Can this last? Are we doing something wrong?" Scientific evidence about the pollution of our cities, including for example the smog in Los Angeles, gave warning signals. The explosion of knowledge in every field and the easy access to that knowledge, through books, print media, and television, has led to a holistic understanding of the workings of nature on this "Spaceship Earth". This availability of knowledge and the serious efforts of dedicated people, including the scientists, ecologists, and environmental activists, to apply that knowledge to balance growth with environmental protection could yet result in a real and wholesome progress of the *Homo sapien*. The collective consciousness of the technical world seems to be compelling towards environmental preservation.

Enter NEPA — the National Environmental Policy Act of 1969. NEPA is a milestone in man's understanding of the relationship between his own survival and the survival of the total ecology that has supported all life on earth, kept its water and air pure, and nurtured the soul-enchanting beauty of its wilderness areas. At last, the technological man realized that nature is an organic whole and no part of nature can be long insulted without damaging the whole. The human being lives in a dynamic environmental system whose efficient functioning has evolved over the ages. Into this system grew the modern man with his powerful industrial intellect which seemed for a while to ignore the long-term effects for the short-term goals of profit and pleasure. The National Environmental Policy Act (NEPA) is the harbinger of the dawn of a new age, signifying both the end of science and the beginning of science. NEPA heralds the end of the myopic era of the application of science for the mere immediate production of goods for comfort, and ushers in the new era of science which asks that industrial civilization develop in conformance with the overall needs of all of nature by understanding the higher dimensions of human experience and destiny. In short, the NEPA statute intends that man be in tune with nature for the good of his own future and the future of his progeny. The purpose of this book is to describe, in practical terms, the application of NEPA to our everyday decision making, so that the harmony of nature with man is achieved and preserved for generations to come.

Lest we rest easy thinking that faithful implementation of NEPA by decision makers is an inexorable promise, the author would warn you, the reader, that decision makers are human and, like the rest of us, first look for the line of least resistance. Tradition, institutionalized expectations, and habit compel the decision maker to place primary emphasis on the immediate need, i.e., success of the project, not on the altruistic goal of environmental protection which seems too unreal, abstract, and futuristic to commit funds or effort in this area, as easily as committing funds for, say, engineering design or cost estimation. The entrenched feeling is that NEPA is a law that must be obeyed, alas, since it cannot be avoided! Most decision makers do not at first appreciate that NEPA can actually help make better decisions. With rare exceptions, this misunderstanding at all levels of management renders the task of the environmental professional team doubly difficult since, instead of getting credit for contributing towards better decisions, the team may be looked upon as a nuisance at best and as peddlers of a soft science. The police officers of the nation can probably relate very well to this feeling. The author has had the good fortune to have met and worked with several of these rare exceptions among decision makers. Nevertheless, they are exceptions, not the rule. So good luck to you, the practicing environmental professional, and take heart because despite the mistrust, the environmental impact statement has often served as a useful, multifaceted decision-making tool for many major projects in America.

II. ORGANIZATION

This book is organized to provide adequate and systematic insight to the implementation of NEPA. The main organizational flow of the book is thus:

Description of environmental statutes (Chapter 2)
↓
Description of the technical aspects of these statutes (Chapter 3)
↓
The NEPA process (the theme of the book, Chapter 4)
↓
How to write environmental documents (the how-to aspects of NEPA)
↓
Selecting control technologies to comply with stautes (Chapter 6)
↓
What to do after completing NEPA compliance? (Chapter 8)
↓
Empathizing with the public (Chapter 9)
↓
Questions and answers for use by NEPA professionals (Chapter 10)

Thus, the book first describes the environmental statutes and the Council on Environmental Quality Regulations to implement NEPA. Then, the application of NEPA is illustrated throughout the book by weaving in the experience of preparing NEPA documentation with an example, a coal-fired power plant. The selection of control technologies is also discussed using this power plant example. The importance of communicating with the public and methods to achieve it are described. The burning topic of the day — the safe disposal of nuclear wastes — is briefly addressed for the readers benefit. The author's hope is that this book provides the reader with sufficient methodological knowledge to implement NEPA so that man and nature can live in productive harmony.

III. SUMMARY

Implementing the NEPA process requires an understanding and application of the principles of balancing the technology-environment-economics triad described in Chapter 2. To understand these principles, the environmental statutes should be understood from their implementation aspects by careful understanding of the technical aspects of these statutes as described in Chapter 3. The heart of the application of the NEPA process during the decision-making phase is an unbiased analysis of the alternatives to a proposed action. The NEPA professional becomes a policy analyst in providing to the decision maker an evaluation of alternatives. Discharging the responsibilities in this role of policy analyst requires an understanding of the ethics and obligations of the professional to the public. This is discussed in Chapter 5 (see guidelines for alternatives analysis). If the NEPA professional performs his task adequately, the NEPA process will help public officials make decisions after a full review and understanding of the environmental consequences. Then only will the process lead to actions that protect, restore, and enhance the environment as intended by the National Environment Policy Act of 1969. The continued viability of this important law depends entirely on the will of the decision makers, the scientific and technical professionals, and of an ever vigilant public, to all of whom this book is dedicated with hope and eager expectation of usefulness.

Chapter 2

THE ENVIRONMENTAL STATUTES

I. INTRODUCTION

Environmental statutes are more simply called environmental laws. The early origins of such laws can be traced to government control over coal burning in the 16th century and regulation of sewage disposal in the 19th century.[3] By the early 1900s, legislation set aside national parks and governed mining and timber activities. The current era of environmental statutes began at the end of the 1960s. By the early 1970s, it was clear that a law here and a law there would not be adequate to protect the environment. The public awareness of the damage to the environment by the immense industrial mammoth created by man led to an environmental statute of deceptive simplicity, the National Environmental Policy Act of 1969 (Public Law No. 91-190, approved January 1, 1970).

The National Environmental Policy Act (NEPA) advocates a balanced triad of technology, environment, and economics schematically shown in Figure 1.

The NEPA could be considered as a paragon of legislations in that it is visionary, effective, and comprehensive, but brief (see Appendix A). Although deceptively simple, it has created a powerful medium for all societal entities to express and enforce their concern for the preservation and enhancement of environmental quality through an orderly process of public participation. NEPA is perhaps the only environmental statute that not only emphasizes the organic relationship between man and his environment, but also directs each agency of the federal government to insure that man's activities are conducted so as to preserve and enhance the environment. To this end, the NEPA makes some very important statements as summarized below:

1. It is the continuing responsibility of the federal government to use all practicable means, consistent with other essential considerations of national policy, to improve and coordinate federal plans, functions, programs, and resources to the end that the nation may

   a. Fulfill the responsibilities of each generation as trustee of the environment for succeeding generations.
   b. Assure for all Americans safe, healthful, productive, and esthetically and culturally pleasing surroundings.
   c. Attain the widest range of beneficial uses of the environment without degradation, risk to health or safety, or other undesirable and unintended consequences.
   d. Preserve important historic, cultural, and natural aspects of our national heritage and maintain, wherever possible, an environment which supports diversity and variety of individual choice.
   e. Achieve a balance between population and resource use which will permit high standards of living and a wide sharing of the amenities of life.
   f. Enhance the quality of renewable resources and approach the maximum attainable recycling of depletable resources.

2. The U.S. Congress recognizes that each person should enjoy a healthful environment and that each person has a responsibility to contribute to the preservation and enhancement of the environment.

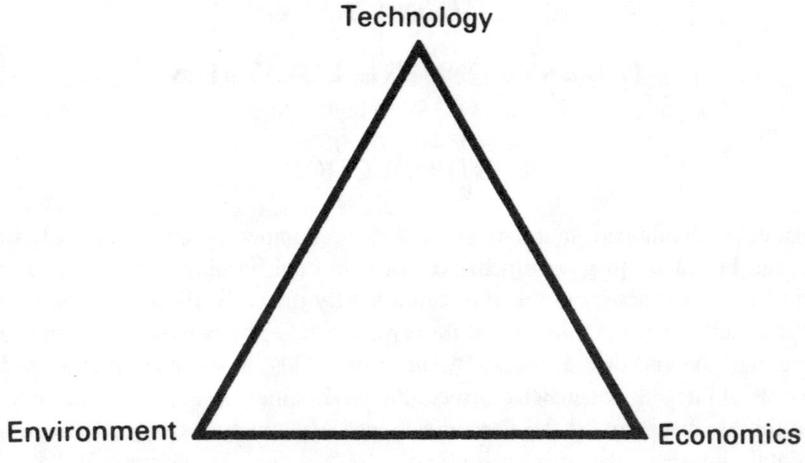

FIGURE 1. NEPA advocates balanced triad of three considerations.

3. The U.S. Congress authorizes and directs that, to the fullest extent possible, all agencies of the federal government shall

   a. Utilize a systematic, interdisciplinary approach which will insure the integrated use of the natural and social sciences and the environmental design arts in planning and in decison making which may have an impact on man's environment.
   b. Identify and develop methods and procedures, in consultation with the Council on Environmental Quality, established by Title II of this Act, which will insure that presently unquantified environmental amenities and values may be given appropriate consideration in decision making along with economic and technical considerations.
   c. Include in every recommendation or report on proposals for legislation and other major federal actions significantly affecting the quality of the human environment, a detailed statement by the responsible official on

      i. The environmental impact of the proposed action.
      ii. Any adverse environmental effects which cannot be avoided should the proposal be implemented.
      iii. Alternatives to the proposed action.
      iv. The relationship between local, short-term uses of man's environment and the maintenance and enhancement of long-term productivity.
      v. Any irreversible and irretrievable commitments of resources which would be involved in the proposed action should it be implemented.

In essence, NEPA recognizes that our home, the ''Spaceship Earth'', has limited ability to cope with abuse. This thought is well put in the 11th annual report of the president's Council on Environmental Quality, as follows:[5]

- We can no longer assume that the earth will heal and renew itself indefinitely.
- Human numbers and activity are catching up with the ability of the earth to recover.
- Careful management of our resources of the planet — its water, soils, minerals, forests, fish, and air — should help sustain us in great numbers.
- The quality of our life will rest on careful stewardship and husbandry of the resources of our planet.

II. ATTITUDES TO NEPA

The author conducted a survey as part of the short course that he teaches under the joint auspices of the University of Washington, Washington State University, and Oregon State University. The survey included the following questions:

1. Has the influence of NEPA on the awareness of environmental problems and their impact on the health and safety of the U.S. population been (a) positive, (b) mildly effective, (c) negligible, or (d) unknown?
2. Has NEPA generally helped avoid expensive mistakes in the execution of major projects in the U.S.?
3. Do you feel the requirements of NEPA are clearly understood for ease of implementation?
4. Has NEPA helped or hindered in executing your projects?
5. Do you think that NEPA should be enforced more vigorously than at present in the U.S.?
6. Do you feel that the environmental documents you have read are (a) too complex, (b) just right, or (c) inadequate?

About 100 responses from scientists, engineers, and project managers to the survey were received. The majority of the respondents felt that the influence of NEPA was positive (52%) and 37% felt it was mildly effective. Again, a majority felt that NEPA has or may have helped avoid expensive mistakes in the execution of major projects in the U.S.

Also, the response clearly indicated that the requirements of NEPA were **not** clearly understood and that the environmental documents were too complex and generally inadequate.

The intent of this book is to help the reader understand NEPA requirements more clearly so that its implementation will be more efficient and effective. The students attending the 2-day short course (on which this book is based) expressed their opinion that the course was well structured to benefit them. It is the intention of the author to structure the contents of this book along the lines of the course so that it benefits project managers, construction managers, project engineers, conceptual design engineers, scientists, ecologists, NEPA document reviewers, planners, and executives responsible for large projects' management. Compliance with the requirements of NEPA is a vital step in insuring successful completion of projects.

III. COUNCIL ON ENVIRONMENTAL QUALITY REGULATIONS FOR IMPLEMENTING NEPA

What exactly are the requirements of NEPA? This question becomes very real and pressing because NEPA itself (presented in Appendix A) does not define the requirements in a way that they can be easily implemented. To remedy this situation, the Council on Environmental Quality (CEQ) which was established by law (NEPA), has published *Regulations for Implementing the Procedural Provisions of The National Environmental Policy Act*. The issuing of these regulations is one of the responsibilities of the CEQ defined in Executive Order 11514, March 5, 1970, amended by Executive Order 11991, May 24, 1977.

The CEQ regulations are published in Title 40 of the Code of Federal Regulations, Parts 1500 to 1508 and cited as 40 CFR 1500 to 1508. These regulations are presented in Appendix B. While all aspects of these regulations about NEPA implementation are important and should be carefully studied, some topics worthy of special highlighting are (1) reducing paperwork, (2) reducing delay, (3) apply NEPA early in the process, (4) lead agency, (5) major actions requiring environmental impact statements, (6) page limits, (7) writing in plain language, (8) incomplete or unavailable information, (9) categorical exclusion, (10) duty to comment, (11) scientific and professional integrity, and (12) NEPA process. These are summarized below from the CEQ regulations.

Reducing paperwork — Reducing paperwork has received a very important treatment in the CEQ regulations (40 CFR 1500.4) because early writers of environmental documents generated literally carloads of documents that nobody could hope to read and grasp. For a while, quantity was considered to be equal to quality. To remedy this problem, CEQ has defined 17 methods of reducing paperwork. The first two are particularly noteworthy:

1. Reduce the length of the documents by setting page limits to the EIS.
2. Prepare analytic, not encyclopedic EISs.

Reducing delay — CEQ has described 12 methods of reducing delay in successful completion of any project from its conception to construction and operation. The very first step one could take to avoid project delays is to integrate the NEPA process into early planning.

Apply NEPA early in the process — Among methods of applying NEPA early, one important method is to comply with the mandate of NEPA to ''utilize a systematic, interdisciplinary approach which will insure the integrated use of the natural and social sciences and the environmental design arts in planning and in decision making which may have an impact on man's environment.''

Lead agency — If more than one federal agency is involved in a project, there shall be a lead agency to supervise the NEPA compliance process.

Major actions requiring environmental impact statements — Federal agencies are required to properly define the criteria for determining the types of proposals (considering size, impacts, costs of project, etc.) which require environmental impact statements, as opposed to environmental documents of lesser detail, complexity, and public participation.

Page limits — CEQ has regulated the length of environmental impact statements to 150 pages of text for proposals of normal scope and to 300 pages for proposals of unusual scope. Although these page limits are regulations, they are more often used as guidelines. The more one adheres to these page limits, the more readable and useful the environmental impact statement.

Writing in plain language — If environmental impact statements should serve the purpose of public participation in the decision-making process, they should be written in plain language and may use appropriate graphics so that decision makers and the public can readily understand them. Agencies should employ writers of clear prose or editors to write, review, or edit statements which will be based upon the analysis and supporting data from the natural and social sciences and the environmental design arts.

Incomplete or unavailable information — This portion of the CEQ regulations has been recently changed. On August 9, 1985, CEQ published a proposed amendment to these regulations (40 CFR 1502.22) which addressed incomplete or unavailable information in an environmental impact statement. After reviewing the comments received in response to the proposal, the CEQ has issued the final amendment to those regulations.

The final amendment requires all federal agencies to disclose the fact of incomplete or unavailable information when evaluating reasonably foreseeable significant adverse impacts on the human environment in an environmental impact statement and to obtain that information if the overall costs of doing so are not exorbitant. If the agency is unable to obtain the information because overall costs are exorbitant or because the means to obtain it are not known, the agency must (1) affirmatively disclose the fact that such information is unavailable; (2) explain the relevance of the unavailable information; (3) summarize the existing credible scientific evidence which is relevant to the evaluation of the agency of significant adverse impacts on the human environment; and (4) evaluate the impacts based upon theoretical approaches or research methods generally accepted in the scientific community. The amendment also specifies that impacts which have a low probability of occurrence, but catastrophic consequences if they do occur, should be evaluated if the analysis is supported by credible scientific evidence, is not based on pure conjecture, and is within the rule of reason. The requirement to prepare a "worst case analysis" is rescinded.

These changes are reflected in the CEQ regulations presented in Appendix B.

Categorical exclusion — "Categorical exclusion" means a category of actions which do not individually or cumulatively have a significant effect on the human environment and which have been found to have no such effect in procedures adopted by a federal agency in implementation of these regulations (§ 1507.3) and for which, therefore, neither an environmental assessment nor an environmental impact statement is required. An agency may decide in its procedures, or otherwise, to prepare environmental assessments for the reasons stated in § 1508.9 even though it is not required to do so.

Duty to comment — Federal agencies with jurisdiction by law or special expertise with respect to any environmental impact involved and agencies which are authorized to develop and enforce environmental standards shall comment on statements within their jurisdiction, expertise, or authority. Agencies shall comment within the time period specified for comment in § 1506.10. A federal agency may reply that it has no comment. If a cooperating agency is satisfied that its views are adequately reflected in the environmental impact statement, it should reply that it has no comment. Examples of federal agencies that have a duty to comment are the U.S. Environmental Protection Agency and the Department of Interior. The type of the proposed project and the environment it affects will determine the federal agencies that should comment. The major executive branch agencies with environmental responsibilities are shown in Figure 2.

Scientific and Professional Integrity — Agencies shall insure the professional integrity, including scientific integrity, of the discussions and analyses in environmental impact statements. They shall identify any methodologies used and shall make explicit reference by footnote to the scientific and other sources relied upon for conclusions in the statement. An agency may place discussion of methodology in an appendix.

The NEPA process — "NEPA process" means all measures necessary for compliance with the requirements of Section 2 and Title I of NEPA. The purpose of the NEPA process is to help federal agencies make better decisions and foster excellent action. The NEPA process should help public officials make decisions after a full review and understanding of the environmental consequences. The process should lead to actions that protect, restore, and enhance the environment. The CEQ regulations (40 CFR 1500 to 1508) constitute the essence of the NEPA process. The process is described in detail in Chapter 4.

IV. OTHER ENVIRONMENTAL LAWS

The National Environmental Policy Act of 1969 is a statute that virtually compels public officials to comply with numerous other statutes that are enacted to protect the environment. In fact, the NEPA process is not complete without evaluating any proposal for compliance with those other environmental laws as appropriate. Some of these laws are listed below:

- The Clean Air Act and amendments thereto
- Federal Water Pollution Control Act, renamed the Clean Water Act in 1977
- Marine Protection, Research, and Sanctuaries Act
- Endangered Species Act
- Noise Control Act
- Resource Conservation and Recovery Act
- Safe Drinking Water Act

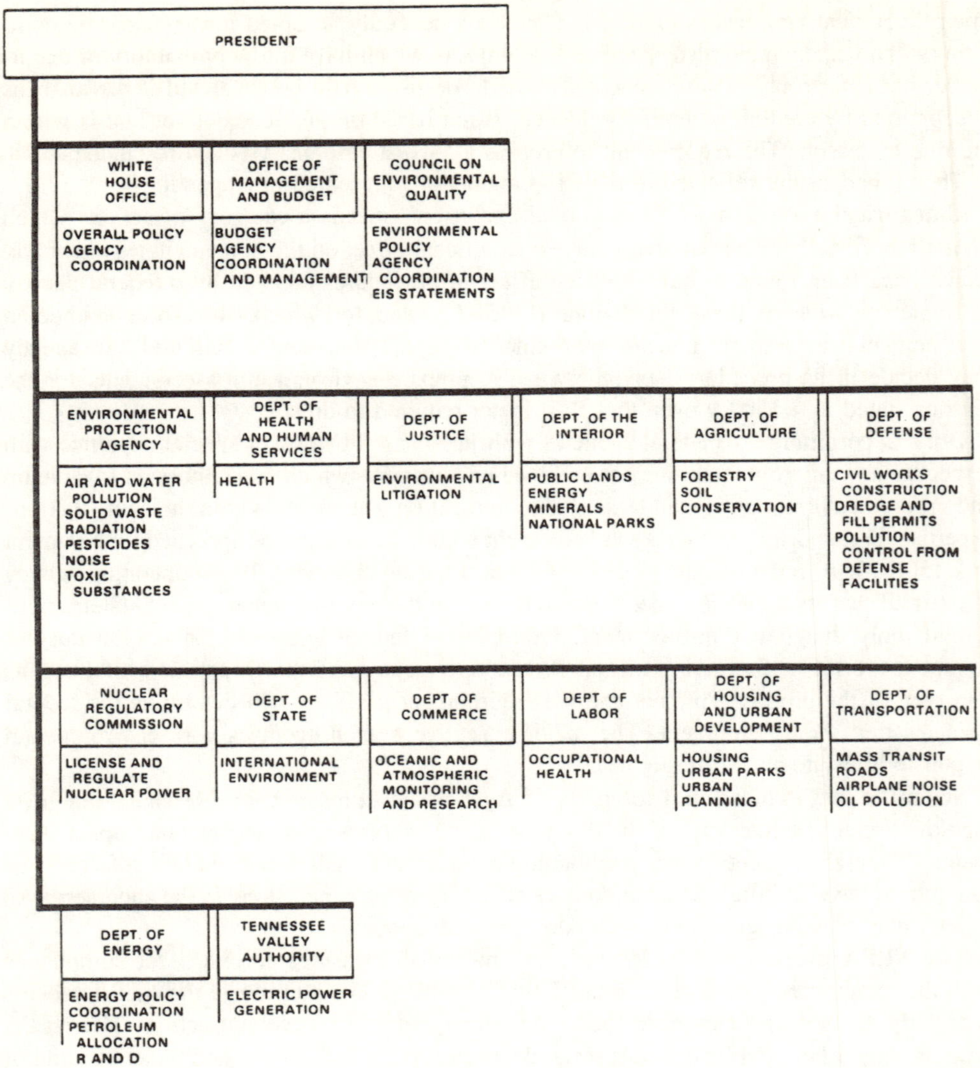

FIGURE 2. Major executive branch agencies with environmental responsibilities.

- Toxic Substances Control Act
- Comprehensive Environmental Response, Compensation, and Liability Act of 1980 (Superfund)

These laws tend to be amended (revised) periodically, with the result that there is confusion about the complete official version at any one time. Also, amendments are contemplated to the statutes as required by subsequent developments in environmental impacts. Therefore, the reader should be careful to obtain the latest status of the laws. A compilation of the laws is published annually by several organizations.[15]

Of these laws, the most important and most frequently used during the conduct of the NEPA process are (1) the Clean Air Act, (2) the Clean Water Act, (3) the Resource Conservation and Recovery Act, and (4) Safe Drinking Water Act. A brief summary of these four acts follow.

A. The Clean Air Act

This act legislates the prevention and control of air pollution to protect and enhance the quality of the air resources of the nation. The law deals with (1) air quality standards, (2) emission control limits, (3) ozone protection, (4) prevention of significant deterioration (PSD), (5) motor vehicle emission standards, (6) aircraft emission standards, (7) air quality control regions, and (8) state implementation plans.

B. The Clean Water Act

The efforts of the U.S. to prevent water pollution can be traced to the 1899 Refuse Act which made it illegal to discharge "refuse matter" into navigable waters. In October 1972, the Federal Water Pollution Control Act was enacted (Public Law 92-500) and subsequently amended by the Clean Water Act of 1977 and is now commonly known as the Clean Water Act.

The objective of this act is to restore and maintain the chemical, physical, and biological integrity of the waters of the nation. To achieve this objective, the act set up the following goals:

1. It is the national goal that the discharge of pollutants into the navigable waters be eliminated by 1985.
2. It is the national goal that wherever attainable, an interim goal of water quality which provides for the protection and propagation of fish, shellfish, and wildlife and provides for recreation in and on the water be achieved by July 1, 1983.
3. It is the national policy that the discharge of toxic pollutants in toxic amounts be prohibited.
4. It is the national policy that federal financial assistance be provided to construct publicly owned waste treatment works.
5. It is the national policy that area-wide waste treatment management planning processes be developed and implemented to assure adequate control of sources of pollutants in each state.
6. It is the national policy that a major research and demonstration effort be made to develop technology necessary to eliminate the discharge of pollutants into the navigable waters, waters of the continuous zone, and the oceans.

It is the policy of the Congress to recognize, preserve, and protect the primary responsibilities and rights of states to prevent, reduce, and eliminate pollution, to plan the development and use (including restoration, preservation, and enhancement) of land and water resources, and to consult with the EPA Administrator in the exercise of his authority under this act.

- Title I of the act provides for research and related programs including development of comprehensive programs for water pollution control.
- Title II of the act provides for federal share grants for construction waste treatment management projects.
- Title III provides for establishing standards and their enforcement including effluent limits from point source discharges, water quality standards, and federal enforcement.
- Title IV describes the permits and licenses aspects of the law including the National Pollutant Discharge Elimination System (NPDES) and permits for disposal of sewage sludge.

C. Resource Conservation and Recovery Act (RCRA)

The objectives of this act are twofold: (1) to protect public health and environment, and

(2) to conserve valuable material and energy resources. These objectives would be attained by:

1. Providing technical and financial assistance to state and local governments and interstate agencies for the development of solid waste management plans (including resource recovery and resource conservation systems) which will promote improved solid waste management techniques (including more effective organizational arrangements), new and improved methods of collection, separation, and recovery of solid waste, and the environmentally safe disposal of nonrecoverable residues.
2. Providing training grants in occupations involving the design, operation, and maintenance of solid waste disposal systems.
3. Prohibiting future open dumping on the land and requiring the conversion of existing open dumps to facilities which do not pose a danger to the environment or to health.
4. Regulating the treatment, storage, transportation, and disposal of hazardous wastes which have adverse effects on health and the environment.
5. Providing for the promulgation of guidelines for solid waste collection, transport, separation, recovery, and disposal practices and systems.
6. Promoting a national research and development program for improved solid waste management and resource conservation techniques, more effective organizational arrangements, new and improved methods of collection, separation, and recovery, recycling of solid wastes, and environmentally safe disposal of nonrecoverable residues;
7. Promoting the demonstration, construction, and application of solid waste management, resource recovery, and resource conservation systems which preserve and enhance the quality of air, water, and resources.
8. Establishing a cooperative effort among the federal, state, and local governments, and private enterprise in order to recover valuable materials and energy from solid waste.

A very significant aspect of this act is the great emphasis it brings to bear on the management of hazardous wastes including standards on the generation, storage, disposal, and transportation of such wastes.

D. Safe Drinking Water Act (SDWA)

This act (Public Law 93-523, December 1974) aims to protect the safety of public water systems. The act directs the EPA administrator to promulgate national interim primary drinking water standards and also to protect the underground sources of drinking water. The law requires the EPA to publish regulations for underground injection control programs.

E. Other Statutes

Other federal statutes that are frequently used during the NEPA process are (1) the National Historic Preservation Act and (2) the Endangered Species Act. All the federal environmental statutes must be reviewed for applicability to the specific proposed action when preparing NEPA documentation for that action. For example, the Noise Control Act and the EPA regulations promulgated under the authority of this act would become important when NEPA documentation is being prepared for, say, construction of a new airport or for using heavy machinery for a major construction activity close to population centers. The environmental law that applies in a situation is a function of the environment that would be impacted and type of environmental impact that would be projected from the action proposed. The projected environmental insult determines which laws should be applicable and cited in the NEPA document.

Chapter 3

TECHNICAL ASPECTS OF STATUTES, STANDARDS, AND REGULATIONS

I. INTRODUCTION

A major goal of the NEPA process is to ensure that decision makers comply with all applicable regulations and standards for a given project. In this sense, NEPA is the umbrella legislation that ensures that all other environmental protection statutes are incorporated, as applicable, into the project design during NEPA implementation. The EIS is specifically structured to achieve this objective as described in Chapter 5. However, to comply with the statutes it is essential to convert their generalized requirements into the technical language of effluent limitation quantities and standards which can be met. The various policing and regulating agencies of the executive branch of the federal government (such as the U.S. Environmental Protection Agency and the U.S. Nuclear Regulatory Commission) have published technical standards for easier compliance with the statutes under the authority of these statutes. This chapter introduces these technical standards as preliminary to the NEPA process implementation discussed in later chapters.

Earlier, in Chapter 2, we touched upon at least ten environmental protection statutes that are in force today. These are

1. The National Environmental Policy Act (NEPA)
2. The Clean Air Act (CAA)
3. The Clean Water Act (CWA)
4. Endangered Species Act (ESA)
5. Noise Control Act (NCA)
6. Resource Conservation and Recovery Act (RCRA) (active sites)
7. Safe Drinking Water Act (SDWA)
8. Toxic Substances Control Act (TOSCA)
9. Comprehensive Environmental Response, Compensation, and Liability Act of 1980 (CERCLA of Superfund) (inactive sites)
10. Marine Protection, Research, and Sanctuaries Act

There are at least 11 more acts that are as equally important in protecting the environment and public health and safety; but these are less frequently visible unlike, for example, the CAA, CWA, NEPA, or RCRA. These are

11. Federal Insecticide, Fungicide, and Rodenticide Act (FIFRA)
12. National Low Level Radioactive Waste Policy Act of 1980
13. Nuclear Waste Policy Act of 1983
14. National Historic Preservation Act
15. Used Oil Recycling Act of 1980
16. Fish and Wildlife Coordination Act (50 CFR, Part 410)
17. Wild and Scenic Rivers Act (16 USC 1271-1287)
18. Navigation and Navigable Waters Act (15 CFR, Part 320)
19. Coastal Zone Management Act (15 CFR, Part 930)
20. Occupational Safety and Health Act (OSHA)
21. The Environmental Quality Improvement Act of 1970

All these statutes, except for CAA perhaps, address environmental protection qualitatively in terms of policy goals, not in quantitative terms that specify limits for emissions or ambient concentrations of pollutants. The statutes provide the necessary authority to the executive branch of the federal government to protect the air, land, and water resources by use of technology deemed most appropriate by the various environmental policy agencies such as the U.S. EPA. The unwillingness of the U.S. Congress to fix forever the technology and emission limits in the statutes is understandable because technology changes with time and periodic adjustment to emission limits may be required in response to various factors including the overall economic climate globally and in the U.S. Some flexibility in achieving the goals of the statutes is a wise step that our lawmakers have allowed.

Other mechanisms which are important in achieving the provisions of the statutes are

- Executive orders
- Federal regulations and standards
- State environmental laws
- State regulations and standards
- Local (county, city) regulations and standards
- Industry self-policing regulations and standards

This chapter will address the technical aspects of the statutes as defined by the various environmental protection and enforcement agencies of the government. Detailed discussion of either the statutes or these other mechanisms would occupy several volumes. What is possible to present in this book is a guide to these statutes and the way standards and regulations are derived from them. The standards and regulations are subject to periodic revisions. Therefore, the careful professional should consult the latest version in the Code of Federal Regulations (CFR) in which they are published annually. Revisions to CFR are published as they occur in the *Federal Register* (FR) published every working day by the federal government office of the *Federal Register*.

II. DESCRIPTION OF THE CODE OF FEDERAL REGULATIONS

The Code of Federal Regulations is a codification of the rules published in the *Federal Register* by the executive departments and agencies of the federal government. The code is divided into 50 titles which represent broad areas subject to federal regulation. Each title is divided into chapters which usually bear the name of the issuing agency. Each chapter is further subdivided into parts covering specific regulatory areas.

A. Issue Dates

Each volume of the Code is revised at least once each calendar year and issued approximately as follows:

- Title 1 through Title 16 .... as of January 1
- Title 17 through Title 27 .... as of April 1
- Title 28 through Title 41 .... as of July 1
- Title 42 through Title 50 .... as of October 1

The appropriate revision date is printed on the cover of each volume.

B. Legal Status of the Code of Federal Regulations

The contents of the *Federal Register* are required to be judicially noticed (44 U.S.C. 1507). The Code of Federal Regulations is prima facie evidence of the text of the original documents (44 U.S.C. 1510).

C. How to Use the Code of Federal Regulations

The Code of Federal Regulations is kept up to date by the individual issues of the *Federal Register*. These two publications must be used together to determine the latest version of any given rule.

To determine whether a code volume has been amended since its revision date, consult the "List of CFR Sections Affected (LSA)", which is issued monthly, and the "Cumulative List of Parts Affected", which appears in the Readers Aids section of the daily *Federal Register*. These two lists will identify the *Federal Register* page number of the latest amendment of any given rule. The environmental professional is usually interested in two of these 50 titles: Titles 10 and 40.

Title 10 covers regulations governing the production and utilization of energy by the NRC and DOE. Title 10 is composed of Volumes I, II, III, X, and XV as follows:

I. Nuclear Regulatory Commission (Parts 0 to 199)
II. Department of Energy (Parts 200 to 699)
III. Department of Energy (Parts 700 to 999)
X. Department of Energy (General Provisions) (Parts 1000 to 1099)
XV. Office of Federal Inspector for the Alaska Natural Gas Transportation System (Parts 1500 to 1599)

The title is published every year with revisions effective January 1.

Title 40 covers the protection of environment and is composed of nine volumes. Each of these nine volumes contain parts arranged in the following order: Parts 0 to 51, Part 52, Parts 53 to 80, Parts 81 to 99, Parts 100 to 149, Parts 150 to 189, Parts 190 to 399, Parts 400 to 424, and Part 425 to the end. The contents of these volumes represent all current regulations codified under this title of the CFR as of July 1 of each year. Chapter I — Environmental Protection Agency appears in all nine volumes. The Pesticide Tolerance Chemical Index appears in Parts 150 to 189. Regulations issued by the Council on Environmental Quality appear in the volume containing Part 524 to the end. Title 40 is published every year with revisions effective July 1.

Executive orders are published in Title 3. Regulations relating to transportation appear in Title 49. Wildlife and Fisheries Regulations are found in Title 50.

III. REVIEW OF SELECTED ENVIRONMENTAL STATUTES

A review of the objectives and requirements of the five most often used statutes is warranted: (1) NEPA, (2) CAA, (3) CWA, (4) RCRA, and (5) SDWA. Of these, NEPA is the umbrella legislation which, through its environmental impact statement requirement derived from Section 102(c), forces the lead agency to show how the agency would comply with all applicable statutes, standards, and regulations. Since NEPA is reproduced in full in Appendix A, it is not summarized here.

A. Objectives of the Clean Air Act

1. To protect and enhance the quality of the national air resources so as to promote the public health and welfare and the productive capacity of its population.
2. To initiate and accelerate a national research and development program to achieve the prevention and control of air pollution.
3. To provide technical and financial assistance to state and local governments in connection with the development and execution of their air pollution prevention and control programs.
4. To encourage and assist the development and operation of regional air pollution control programs.

The requirements of the CAA are numerous and only a few are listed, as follows:

1. To conduct cooperative activities with the states and, as far as possible, enact uniform standards relating to prevention and control of air pollution.
2. To conduct research, investigation, training, and other activities.
3. To conduct research relating to fuel and vehicles.
4. To provide grants to state and local air pollution control agencies for support of air pollution planning and control programs.
5. To provide grants to interstate air quality agencies or commissions.
6. To set up air quality control regions (AQCRs).
7. To encourage states to submit state implementation plans (SIPs) specifying the manner in which the ambient air quality standards will be achieved in each AQCR.
8. To publish air quality criteria and control technique documents.
9. To promulgate national ambient air quality standards (NAAQS).
10. To develop standards of performance for new stationary sources.
11. To promulgate national emission standards for hazardous air pollutants (NESHAP).
12. To develop federal enforcement procedures including inspection, monitoring, and entry.
13. To prevent U.S. contribution to international air pollution when required.
14. To control air pollution from federal facilities.
15. To assess and collect noncompliance penalty.
16. To take measures to prevent economic disruption and unemployment from pollution prevention efforts.
17. To prevent interstate pollution.
18. To provide for ozone protection in stratosphere.
19. To prevent significant deterioration of air quality (PSD) by promulgating increments and ceilings.
20. To define requirements for nonattainment areas of the nation.
21. To promulgate motor vehicle emission and fuel standards and enforcement procedures.
22. To establish and enforce aircraft emission standards.
23. To conduct detailed and comprehensive studies of the cost of implementing programs necessitated by CAA, this including an economic impact assessment.
24. To establish an air quality monitoring system throughout the U.S.
25. To establish standardized air quality modeling.

The Clean Air Act has not only helped to improve the air quality of the U.S. significantly since 1970, but also has created a new technology and a new science which employs more than 200,000 professionals and technicians.

B. Objectives of the Clean Water Act

The objective of the Clean Water Act is to restore and maintain the chemical, physical, and biological integrity of the national waters. In order to achieve this objective, the act declares that

1. It is the national goal that the discharge of pollutants into the navigable waters be eliminated by 1985.
2. It is the national goal that wherever attainable, an interim goal of water quality which provides for the protection and propagation of fish, shellfish, and wildlife and provides for recreation in and on the water be achieved by July 1, 1983.
3. It is the national policy that the discharge of toxic pollutants in toxic amounts be prohibited.

4. It is the national policy that federal financial assistance be provided to construct publicly owned waste treatment works.
5. It is the national policy that areawide waste treatment management planning processes be developed and implemented to assure adequate control of sources of pollutants in each state.
6. It is the national policy that a major research and demonstration effort be made to develop technology necessary to eliminate the discharge of pollutants into the navigable waters, waters of the contiguous zone, and the oceans.

The act has four titles, as follows:

Title I. Research and related program requiring among other things: (1) development of a comprehensive program for water pollution control in cooperation with federal, state, interstate, municipal, and industry water pollution control organizations and departments and (2) establish a national program for the prevention, reduction, and elimination of water pollution through research, investigations, training, and information.

Title II. Grants and construction of treatment works encouraging waste treatment which results in the construction of revenue-producing facilities providing for (1) the recycling of potential sewage pollutants through the production of agriculture, silviculture, or aquaculture products, or any combination thereof; (2) the confined and contained disposal of pollutants not recycled; (3) the reclamation of wastewater; and (4) the ultimate disposal of sludge in a manner that will not result in environmental hazards.

Title III. Establishing standards and enforcement which include effluent limitations and national standards of performance for control of discharge of pollutants by use of best-available-demonstrated control technology, processes, operating methods, or other alternatives including, where practicable, no discharge of pollutants.

Title IV. Permits and licenses, including Section 402 National Pollutant Discharge Elimination System (NPDES) and Section 404 permits for discharge of dredged or fill material into the navigable waters at specified disposal sites.

C. Objectives of the Resource Conservation and Recovery Act

The objectives of the Resource Conservation and Recovery Act (RCRA) are to promote the protection of health and the environment and to conserve valuable material and energy resources by

1. Providing technical and financial assistance to state and local governments and interstate agencies for the development of solid waste management plans (including resource recovery and resource conservation systems) which will promote improved solid waste management techniques (including more effective organizational arrangements), new and improved methods of collection, separation, and recovery of solid waste, and the environmentally safe disposal of nonrecoverable residues.
2. Providing training grants in occupations involving the design, operation, and maintenance of solid waste disposal systems.
3. Prohibiting future open dumping on the land and requiring the conversion of existing open dumps to facilities which do not pose a danger to the environment or to health.
4. Assuring that hazardous waste management practices are conducted in a manner which protects human health and the environment.
5. Requiring that hazardous waste be properly managed in the first instance thereby reducing the need for corrective action at a future date.

6. Minimizing the generation of hazardous waste and the land disposal of hazardous waste by encouraging process substitution, materials recovery, properly conducted recycling and reuse, and treatment.
7. Establishing a viable federal-state partnership to carry out the purposes of this act and insuring that the administrator will, in carrying out the provisions of Subtitle C of this act, give a high priority to assisting and cooperating with states in obtaining full authorization of state programs under Subtitle C.
8. Providing for the promulgation of guidelines for solid waste collection, transport, separation, recovery, and disposal practices and systems.
9. Promoting a national research and development program for improved solid waste management and resource conservation techniques, more effective organizational arrangements, and new and improved methods of collection, separation, and recovery, and recycling of solid wastes and environmentally safe disposal of nonrecoverable residues.
10. Promoting the demonstration, construction, and application of solid waste management, resource recovery, and resource conservation systems which preserve and enhance the quality of air, water, and land resources.
11. Establishing a cooperative effort among the federal, state, and local governments and private enterprise in order to recover valuable materials and energy from solid waste.

The U.S. Congress declares it to be the national policy that, wherever feasible, the generation of hazardous waste is to be reduced or eliminated as expeditiously as possible. Waste that is nevertheless generated should be treated, stored, or disposed of so as to minimize the present and future threat to human health and the environment.

The requirements of RCRA are several, some of which are listed below:

1. To provide grants for disposal of discarded tires.
2. The labeling of used lubricating oils.
3. To provide detailed procedures for identification and listing of hazardous waste.
4. To establish standards applicable to generators of hazardous waste.
5. To establish standards for transport of hazardous waste.
6. To establish standards applicable to owners and operators of hazardous waste treatment, storage, and disposal facilities.
7. To review the disposal into deep injection wells of waste, solvents, and dioxins.
8. To provide variances from land disposal prohibitions.
9. To provide permits for treatment, storage, or disposal of hazardous wastes.
10. To inspect hazardous waste facilities.
11. To prepare hazardous waste site inventory and conduct monitoring, analysis, and testing.
12. To regulate leaking underground storage tanks (LUST).

D. Objectives of the Safe Drinking Water Act

The objective of the Safe Drinking Water Act is to insure that the drinking water used by any person in the U.S. is safe and will not injure the health and safety of the person.

The requirements of the act are numerous and some are listed as follows:

1. To publish national primary drinking water regulations.
2. To establish maximum containment levels for each contaminant which may have adverse effect on the health of persons.
3. To publish regulations for state underground injection control programs.
4. To establish research, technical assistance, information, and training of personnel.

5. To conduct research, studies, and demonstrations relating to the causes, diagnosis, treatment, control, and prevention of physical and mental diseases and other impairments of man resulting directly or indirectly from contaminants in water, or to the provision of a dependably safe supply of drinking water, including
   a. Improved methods (i) to identify and measure the existence of contaminants in drinking water (including methods which may be used by state and local health and water officials) and (ii) to identify the source of such contaminants.
   b. Improved methods to identify and measure the health effects of contaminants in drinking water.
   c. New methods of treating raw water to prepare it for drinking, so as to improve the efficiency of water treatment and to remove contaminants from water.
   d. Improved methods for providing a dependably safe supply of drinking water, including improvements in water purification and distribution, and methods of assessing the health-related hazards of drinking water.
   e. Improved methods of protecting underground water sources of public water systems from contamination.
6. To the maximum extent feasible, provide technical assistance to the states and municipalities in the establishment and administration of public water system supervision programs (as defined in Section 1443(c)(1)).
7. To establish a National Drinking Water Advisory Council.
8. To promulgate standards for bottled drinking water or provide reasons for not promulgating the standards.

It is important to remember that the drinking water standards apply to ''public water systems'' which are defined as systems that regularly supply water to 15 or more connections or to 25 or more persons at least 60 days a year. The standards apply to faucet water and do not apply to rivers, lakes, and aquifers directly. The standards also apply to individual and commercial establishments which supply water to employees or visitors, or customers, or all of these. The standards do not apply to public water systems including organizations:

1. Which consist only of distribution and storage facilities (and do not have any collection and treatment facilities).
2. Which contain all of its water from, but is not owned or operated by, a public water system to which such regulations apply.
3. Which do not sell water to any person.
4. Which are not carriers which convey passengers in interstate commerce.

IV. TECHNICAL ASPECTS OF STATUTES AND REGULATIONS

The U.S. EPA has the statutory responsibility of translating the objectives and requirements of the various environmental protection statutes to the real world by giving an implementable technical definition to the requirements. The U.S. EPA has published (in the 40 CFR series) the standards and regulations by which the intent of the act is to be carried out.

A. The U.S. EPA

The EPA was formally established as an independent agency in the executive branch on December 2, 1970 by President Nixon. It administers several comprehensive environmental protection laws: CAA, CWA, NCA, RCRA, SDWA, CERCLA, TOSCA, FIFRA, and a combination of acts for radiation protection.

The major program offices of the EPA implement and enforce laws, perform research and development, conduct value impact analyses of proposed regulations and standards, and

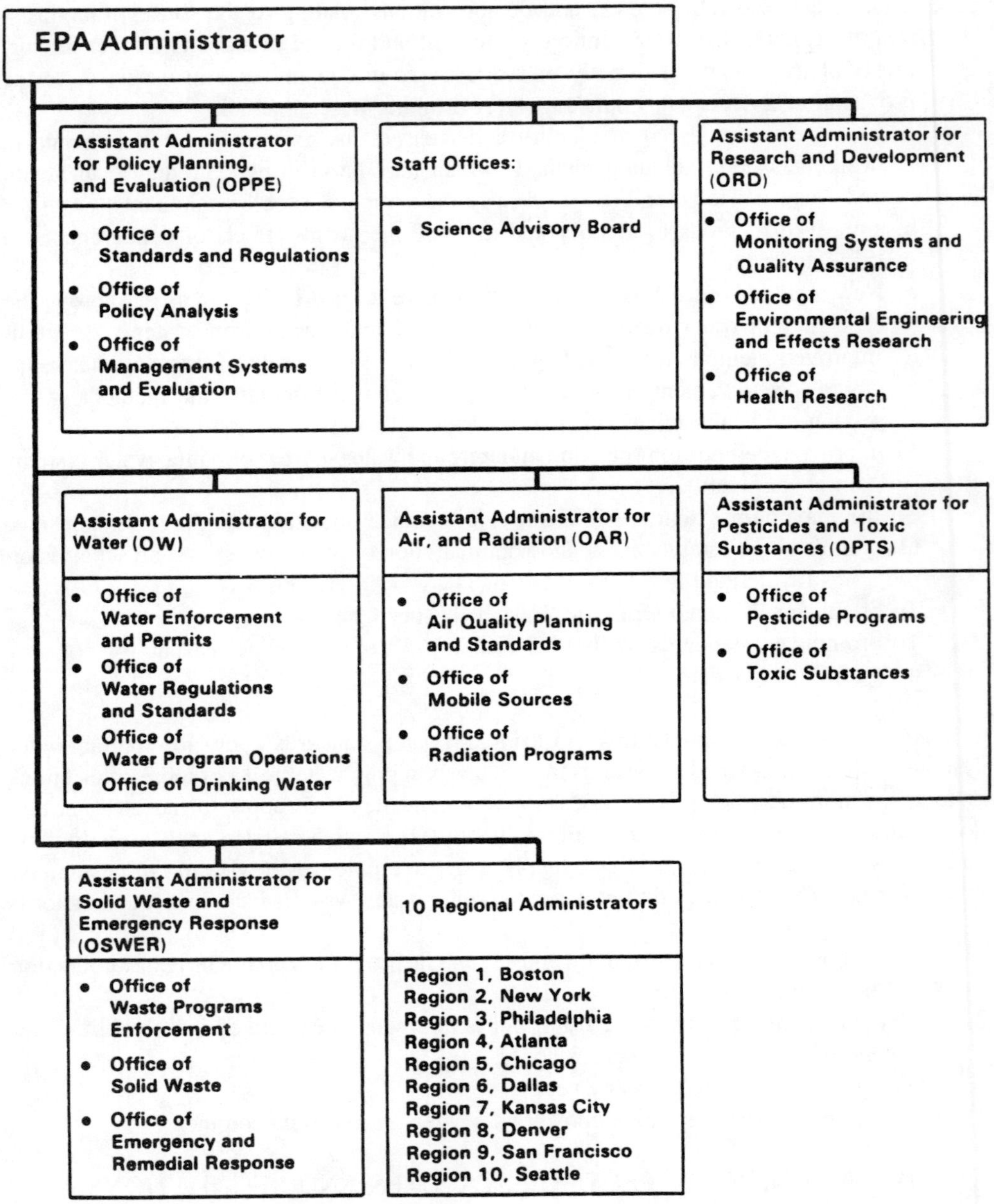

FIGURE 1. Organizational structure of the U.S. EPA.

assess compliance of other federal departments and agencies with the laws administered by EPA through the EIS as the tool. An approximate organizational structure of the EPA is shown in Figure 1.

B. The U.S. EPA Rule Making Process

The EPA follows a formally established rule making process for each regulation and standard it promulgates under the authority of the various environmental statutes. The rule making begins with a start action request (SAR) and ends with the publication of the rule (regulation or standard) in the *Federal Register* as shown in Table 1. The final rule is then codified in CFR titles.

Table 1
EPA RULE MAKING PROCESS

Requirement	Major rule	Significant rule	Minor rule
Start action request (SAR)	Yes	Yes	Yes, unless specially exempt
Development plant	Yes, with AA/OPPE approval	Yes, with SCC[a] approval	No
Options review	Level I or II process	Level I or II process	Not formally needed
Analytical requirements	Regulatory impact analysis	No RIA, but substantial	Minimal or generic
Steering Committee review	Yes	Yes	No
Red border[b]	Yes	Yes	Yes, unless exempt
OMB review[c]	Yes 60 days (proposal) 30 days (final)	Yes 10 days	Yes 10 days unless exempt
Federal Register notification and public comments	Yes	Yes	No

[a] Steering Committee Chairman (SCC). The committee has about 20 members representing nine assistant administrators, ten regional administrators, and an Office of General Counsel. The Office of Standards and Regulations director is the chairperson.

[b] Formal senior management review of all decision packages, normally by assistant administrator/Office of Policy Planning and Evaluation (OPPE) and General Counsel.

[c] OMB = Office of Management and Budget, Executive Office of the President.

The SAR is initiated by a lead office of the EPA (example, the Office of Air and Radiation Programs). The SAR must:

1. Describe the proposed action (rule) and its purpose.
2. State the statutory authority for the action.
3. Propose a classification for the action, i.e., major, significant, or minor.
4. Provide reasons for initiating action and consequences of no action.
5. Certify that the relation of the action to other programs (conflicts, duplication, possibilities for integration) have been investigated; justify unresolved issues.
6. Explain why developing the action now will be an appropriate use of the time and resources of the EPA.

The SAR has to be approved by the assistant administrator (OPPE) in 3 weeks (normally) including a week for review by steering committee members. Once the SAR has been approved, the analytical requirements have to be satisfied. For example, if emission standards setting for a major industry is the action, the analytical requirements will consist of the following activities:

1. Examine alternative standards for emissions.
2. Examine ambient standards (alternative levels).
3. Provide industry description and economic future.
4. Describe a model industry or source being controlled.
5. Analyze model site for benefits and costs of control.
6. Describe pathways through which emissions reach the human body and tissues.
7. Describe potential effects on local, regional, and national populations.
8. Review emission control measures for the operating period of the facility being controlled.
9. Describe the effectiveness and costs of these systems (costs/benefits).
10. Provide cost estimates for representing control methods and efficacy of control.

The findings of the analytical requirements study would eventually become part of either a criteria document used in setting ambient standards or an industry characterization document useful in setting emission standards. The following documents are published by the EPA as part of major rulemaking process under the Clean Air and Clean Water Acts:

When setting ambient air quality or other standards	When setting source performance standards
• Criteria documents based on effects studies	• Industry characterization documents
	• Control technique documents based on model plant for designing control technology

The criteria document for SO_2 standards published in 1969 by the EPA contained detailed information about the various technical items shown in Table 2. Also, during the course of setting standards, the EPA may at its discretion publish an EIS on the proposed action (i.e., the proposed standards) before the standards become final.

Using these procedures for setting standards, the EPA has published final standards of performance for more than 30 industrial categories under the authority of Section III of CAA. These new source performance standards (NSPS) limit emissions to air from these industries, based on the application of the best technological system of continuous emission reduction. Some of the industries for which the NSPS is set in 40 CFR 60 are as follows:

Example CFR citation	Industry
40 CFR 60.40	Fossil fuel-fired steam generators for which construction commenced after August 17, 1971
40 CFR 60.40a	Electric utility steam generating units for which construction commenced after September 18, 1978
40 CFR 60.50	Incinerators
40 CFR 60.60	Portland cement plants
40 CFR 60.70	Nitric acid plants
40 CFR 60.80	Sulfuric acid plants
40 CFR 60.90	Asphalt concrete plants
40 CFR 60.100	Petroleum refineries
40 CFR 60.110	Storage vessels for petroleum liquids
40 CFR 10.120	Secondary lead smelters
40 CFR 10.130	Secondary brass and bronze ingot plants
40 CFR 60.140	Iron and steel plants

To understand how the U.S. EPA has translated the CAA goals to technical terms that are easy of implementation, we can use the example of the coal-fired power plant (from Chapter 5) and illustrate the application of CAA to this plant with the NSPS promulgated in 40 CFR 60.40a through 40 CFR 60.49a, under Subpart Da, which has the following elements:

Subpart Da — Standards of Performance for Electric Utility Steam Generating Units for Which Construction is Commenced After September 18, 1978

60.40a Applicability and designation of affected facility
60.41a Definitions

Table 2
OUTLINE OF AIR QUALITY CRITERIA DOCUMENT FOR SO_2

Properties of pollution

- Concentration
- Chemical composition
- Mineralogical structure
- Absorbed gases
- Coexisting pollutants
- Physical state of pollutant
  - Solid
  - Liquid
  - Gaseous
- Rate of transfer to receptor domain

Measurement Methods

- Hi-vol sampler
- Spot tape sampler
- Dust fall bucket (rate of deposition)
- Condensation nuclei counter
- Impinger (liquid filled)
- Cascade impactor
- Electrostatic precipitator
- Light scattering meter
- Chemical analysis
- Gas analysis (nonadsorbing)
- Absorbed gas analysis
- Light scattering or attenuation (Ringelmann or visibility observation)
- Colored suspension
- Nucleation of precipitation
- Stabilization of fog
- Odor
- Taste

Exposure parameters

- Duration
- Concomitant conditions, such as
  - Temperature
  - Pressure
  - Humidity

Characteristics of receptor

- Physical characteristics
- Individual susceptibility
- State of health
- Rate and site of transfer to receptor

Responses

- Effects on health (diagnosable effects, latent effects, and effects predisposing the organisms to disease)
  - Human health
  - Animal health
  - Plant health
- Effect on human comfort
- Soiling
- Other objectionable surface deposition
- Corrosion of materials
- Deterioration of materials
- Effects on atmospheric properties
- Effects on radiation and temperature

From Air Quality Criteria for Sulfur Oxides, Publ. No. AP-50, U.S. Department of Health, Education, and Welfare, Washington, D.C., 1969.

60.42a Standard for particulate matter
60.43a Standard for sulfur dioxide
60.44a Standard for nitrogen oxides
60.45a Commercial demonstration permit
60.46a Compliance provisions
60.47a Emission monitoring
60.48a Compliance determination procedures and methods
60.49a Reporting requirements

40 CFR 60.40a defines the applicability and designation of affected facility as each electric utility steam generating unit (1) that is capable of combusting more than 73 MW (250 million Btu/hr) heat input of fossil fuel (either alone or in combination with any other fuel); and (2) for which construction or modification is commenced after September 18, 1978. Since

our power plant example from Chapter 5 burns coal (a fossil fuel) and satisfies both these conditions, this part applies to the facility.

Coal contains three major sources of pollutants covered in this regulation: sulfur (S), nitrogen (N), and minerals that become ash when combusted in air. In the combustion zone, pollutants are formed as follows:

$$\text{S (from coal)} + O_2 \text{ (from air)} \rightarrow SO_2 \text{ (in combustion gas)}$$
$$\text{N (from coal)} + O_2 \text{ (from air)} \rightarrow NO_x \text{ (in combustion gas)}$$

The first chemical equation says that one atom of S (atomic weight 32) combines with one molecule of oxygen (mol wt 32) to produce one molecule of SO_2 (mol wt 64). According to the second equation, nitrogen (N) combines with oxygen (O_2) to form nitrogen oxides (NO_x) which is commonly understood to include nitric oxide (NO) and nitrogen dioxide (NO_2).

In our example, 100 lb of coal contains a maximum of 2.3 lb sulfur, 20 lb ash, and generates 1,080,000 Btu of heat energy when burned. This information can be used to evaluate the SO_2 standards given in 40 CFR 60.43a which says, among other things, that no owner or operator subject to the provisions of this subpart shall cause to be discharged into the atmosphere from any affected facility which combusts solid fuel or solid-derived fuel, any gases which contain sulfur dioxide in excess of 520 ng/J (1.20 lb/million Btu) heat input and 10% of the potential combustion concentration (90% reduction). Accordingly, two limitations to emissions apply:

1. Limit emission to 1.2 lb $SO_2/10^6$ Btu.
2. Reduce the SO_2 emission by 90%. This second requirement was modified on July 11, 1979 (FR Volume 44, Part 33580) to state that when burning intermediate or low-sulfur coals, reductions between 70 and 90% are permitted, provided emissions are below 0.6 lb SO_2/million Btu.

In the example power plant, 1 million Btu heat input to the boiler is provided by 92.6 lb of coal which generates 4.26 lb SO_2. To meet the 1.2-lb/million-Btu emission limit for SO_2 only 72% removal of SO_2 is necessary. But to meet the 0.6-lb/SO_2 limit, about 85% SO_2 removal is needed. These calculations are based on the high coal sulfur content of 2.3%, not on the average 1.8% sulfur content coal. Therefore, the power plant need not control SO_2 more than 85%, so far as NSPS requirements are concerned.

It is also necessary to evaluate the required degree of SO_2 control from the viewpoint of the ambient air quality standards and PSD requirements, and control the SO_2 accordingly to meet any higher requirement. The reader will note that in Chapter 5 it is pointed out that SO_2 removal from two of the units would be 85%, and 90% from the other two units to meet the PSD requirements.

Similar procedures apply to evaluating the needed control efficiency of the equipment for particulate removal and for NO_x reduction.

After the plant begins to operate, compliance with the requirements of NSPS is audited using emission monitoring (40 CFR 60.47a) and reporting (40 CFR 60.49a) requirements.

C. National Ambient Air Quality Standards (NAAQS)

Without exception, all documents prepared under the NEPA process use the NAAQS as the reference to which estimated air quality impacts of a project are compared for compliance. Therefore, it is important to understand these standards. According to Section 109 of the CAA, the EPA has promulgated national primary and second ambient air quality standards in 40 CFR 50.

National primary ambient air quality standards define levels of air quality which the administration judges are necessary, with an adequate margin of safety, to protect the public health. National secondary ambient air quality standards define levels of air quality which the administrator judges necessary to protect the public welfare from any known or anticipated adverse effects of a pollutant. Such standards are subject to revision, and additional primary and secondary standards may be promulgated as the administrator deems necessary to protect the public health and welfare.

With reference to PSD increments, the promulgation of national primary and secondary ambient air quality standards shall not be considered in any manner to allow significant deterioration of existing air quality in any portion of any state.

The proposal, promulgation, or revision of national primary and secondary ambient air quality standards shall not prohibit any state from establishing ambient air quality standards for that state or any portion thereof which are more stringent than the national standards.

The reference conditions for measuring the air quality are (1) reference temperature of 25°C and (2) reference pressure of 760 mm of mercury (1013.2 mbar). A summary of these standards published by the U.S. EPA in textual form in 40 CFR 50 is as follows in the text.

1. **40 CFR 50.4** *(National Primary AAQS for Sulfur Oxides)*

The national primary ambient air quality standards for sulfur oxides (sulfur dioxide) measured as sulfur dioxide by the reference method described in Appendix A to this part, or by an equivalent method, are

(a) 80 $\mu g/m^3$ (0.03 ppm) — annual arithmetic mean.
(b) 365 $\mu g/m^3$ (0.14 ppm) — maximum 24-hr concentration not to be exceeded more than once per year.

2. **40 CFR 50.5** *(National Secondary AAQS for Sulfur Oxides)*

The national secondary ambient air quality standard for sulfur oxide (sulfur dioxide) measured as sulfur dioxide by the reference method described in Appendix A* of this part, or by any equivalent method, is 1300 $\mu g/m^3$ (0.5 ppm) maximum 3-hr concentration not to be exceeded more than once per year.

3. **40 CFR 50.6** *(National Primary AAQS for Particulate Matter)*

The national primary ambient air quality standards for particulate matter, measured by the reference method described in Appendix B to this part, or by an equivalent method, are

(a) 75 $\mu g/m^3$ — annual geometric mean.
(b) 260 $\mu g/m^3$ (0.14 ppm) — maximum 24-hr concentration not to be exceeded more than once per year.

4. **40 CFR 50.7** *(National Secondary AAQS for Particulate Matter)*

The national secondary ambient air quality standards for particulate matter, measured by the reference method described in Appendix B to this part, or by an equivalent method, are

(a) 60 $\mu g/m^3$ — annual geometric mean, as a guide to be used in assessing implementation plans to achieve the 24-hr standard.
(b) 150 $\mu g/m^3$ — maximum 24-hr concentration not to be exceeded more than once per year.

* The reader can find these appendixes in the 40 CFR 50 title.

5. **40 CFR 50.8** (*National Primary and Secondary AAQS for Carbon Monoxide*)

The national primary and secondary ambient air quality standards for carbon monoxide, measured by a reference method based on Appendix C to this part and designated in accordance with Part 53 of this chapter, or by an equivalent method, are

(a) 10 mg/m^3 (9 ppm) — maximum 8-hr concentration not to be exceeded more than once per year.
(b) 40 mg/m^3 (35 ppm) — maximum 1-hr concentration not to be exceeded more than once per year.

6. **40 CFR 50.9** (*National Primary and Secondary AAQS for Ozone*)

(a) The level of the national primary and secondary ambient air quality standards for ozone measured by a reference method based on Appendix D to this part and designated in accordance with Part 53 of this chapter, is 0.12 ppm (235 μg/m^3). The standard is attained when the expected number of days per calendar year with maximum hourly average concentration above 0.12 ppm (235 μg/m^3) is equal to or less than 1, as determined by Appendix H.

7. **40 CFR 50.11** (*National Primary and Secondary AAQS for Nitrogen Dioxide*)

(a) The level of the national primary ambient air quality standard for nitrogen dioxide is 0.053 ppm (100 μg/m^3), annual arithmetic mean concentration.
(b) The level of national secondary ambient air quality standard for nitrogen dioxide is 0.053 ppm (100 μg/m^3), annual arithmetic mean concentration.
(c) The levels of the standards shall be measured by
 (1) A reference method based on Appendix F and designated in accordance with Part 53 of this chapter.
 (2) An equivalent method designated in accordance with Part 53 of this chapter.
(d) The standards are attained when the annual arithmetic mean concentration in a calendar year is less than or equal to 0.053 ppm, rounded to three decimal places (fractional parts equal to or greater than 0.0005 ppm must be rounded up). To demonstrate attainment, an annual mean must be based upon hourly data that are at least 75% complete or upon data derived from manual methods that are at least 75% complete for the scheduled sampling days in each calendar quarter.

8. **40 CFR 50.12** (*National Primary and Secondary AAQS for Lead*)

National primary and secondary ambient air quality standards for lead and its compounds, measured as elemental lead by a reference method based on Appendix G to this part, or by an equivalent method, are 1.5 μg/m^3, maximum arithmetic mean averaged over a calendar quarter.

The standards presented above in text form may be somewhat cumbersome to follow. Therefore, I have presented these standards in table format in Table 3.[6] The table is less precise in the definition of the standards, although the numbers are the same as the text. For a clear understanding of the standards, refer to both the text and table.

D. Prevention of Significant Deterioration

Compliance with the NAAQS shall not be considered sufficient and is not an implicit permission to allow significant deterioration of air quality in any portion of any state. Accordingly, 40 CFR 51.24 defines the methods by which the prevention of significant deterioration of air quality can be achieved. A PSD increment is defined for particulate

Table 3
NATIONAL AMBIENT AIR QUALITY STANDARDS (NAAQS)

Pollutant	Primary (health related)		Secondary (welfare related)	
	Averaging time	Concentration	Averaging time	Concentration
TSP	Annual geometric mean	75 $\mu g/m^3$	Annual geometric mean	60 $\mu g/m^3$ [a]
	24 hr	260 $\mu g/m^3$	24 hr	150 $\mu g/m^3$
SO_2	Annual geometric mean	(0.03 ppm) 80 $\mu g/m^3$	3-hr geometric mean	(0.50 ppm) 1300 $\mu g/m^3$
	24 hr	(0.14 ppm) 365 $\mu g/m^3$)		
CO	8 hr	(9 ppm) 10 mg/m^3	Same as primary	
	1 hr	(35 ppm) 40 mg/m^3	Same as primary	
SO_2	Annual geometric mean	(0.053 ppm) 100 $\mu g/m^3$	Same as primary	
O_3	Maximum daily 1-hr average	(0.12 ppm) 235 $\mu g/m^3$	Same as primary	
Pb	Maximum quarterly average	1.5 $\mu g/m^3$	Same as primary	

Note: The standards are categorized for long- and short-term exposure. Long-term standards specify an annual or quarterly mean that may not be exceeded; short-term standards specify upper limit value for 1-, 3-, 8-, and 24-hr averages. The short-term standards are not to be exceeded more than once per year. For example, the ozone standard requires that the expected number of days per calendar year with daily maximum hourly concentrations exceeding 0.12 ppm be less than or equal to one.

[a] This annual geometric mean is a guide used in assessing implementation plans to achieve the 24-hr standards of 150 $\mu g/m^3$.

matter and sulfur dioxide which sets limits on the maximum allowable increase in ambient air concentration even if the NAAQS are fully met. The increments vary for each area defined as follows:

- Class I areas include certain national parks and wilderness areas where the air will continue to be maintained in its pristine state by effectively precluding any substantial industrial growth. For example, all of the following areas which were in existence on August 7, 1977 shall be Class I areas and may not be redesignated:

    1. International parks
    2. National wilderness areas which exceed 5000 acres in size
    3. National memorial parks which exceed 5000 acres in size
    4. National parks which exceed 6000 acres in size

- Class II areas include most areas of the country in which moderate growth is allowed.
- Class III areas include heavily industrialized areas where growth is allowed up to the levels of the secondary NAAQS.

Ambient air concentration increases will be limited by use of emission limitations and such other measures as necessary to assure that in areas designated as Class I, II, or III, increases in pollutant concentration over the baseline concentration shall be limited to the following:

Pollutant	Maximum allowable increase (μg/m³)
Class I	
Particulate matter	
Annual geometric mean	5
24-hr maximum	10
Sulfur dioxide	
Annual geometric mean	2
24-hr maximum	5
3-hr maximum	25
Class II	
Particulate matter	
Annual geometric mean	19
24-hr maximum	37
Sulfur dioxide	
Annual geometric mean	20
24-hr maximum	91
3-hr maximum	512
Class III	
Particulate matter	
Annual geometric mean	37
24-hr maximum	75
Sulfur dioxide	
Annual geometric mean	40
24-hr maximum	182
3-hr maximum	700

1. Ambient Air Ceilings

The plan shall provide that no concentration of a pollutant shall exceed:

1. The concentration permitted under the national secondary ambient air quality standards.
2. The concentration permitted under the national primary ambient air quality standard, whichever concentration is lowest for the pollutant for a period of exposure.

This is a brief summary of the PSD plan requirements described in detail in 40 CFR 51.24. This should be viewed as general information only since there are numerous details, exceptions, and class area redesignation possibilities defined in the regulations.

E. Water Quality Standards (WQS)

Similar to the NAAQS described so far, the knowledge and use of water quality standards and the NPDES permit process is a vital aspect of compliance with the NEPA process since most proposals have the potential to impact the quality of the waters of the U.S. in addition to air quality impacts.

1. NPDES Permit

The NPDES regulations cover basic EPA permitting requirements (40 CFR 122). The regulations also describe what a state must do to obtain approval to operate its own program in lieu of a federal program (40 CFR 123).

The scope of the NPDES permit requirement includes permits for the discharge of pollutants from any point source into the waters of the U.S. (formerly called ''navigable waters of the United States''). Further details of the permit process can be found in 40 CFR 122, 123, 124, and 125. The toxic pollutant effluents are subject to the regulations under 40 CFR 129.

2. Drinking Water Standards

The U.S. EPA has promulgated, in 40 CFR 141, the National Primary Drinking Water Regulations. These regulations, administered by the EPA, implement the Public Health Service Act as amended by the Safe Drinking Water Act, and establish standards applicable to community water systems (not to groundwaters*). ''Community water system'' means a public water system which serves at least 15 service connections used by year-round residents or regularly serves at least 25 year-round residents. Noncommunity water system means a public water system that is not a community water system defined previously.

The 40 CFR 141 regulations promulgate maximum contaminant levels for (1) inorganic chemicals (40 CFR 141.11), (2) organic chemicals (40 CFR 131.12), (3) turbidity (40 CFR 141.13), (4) microbiological contaminants (40 CFR 141.14), (5) radium-226, radium-228, and gross alpha particle radioactivity (40 CFR 141.15), and (6) beta particle and photon radioactivity from man-made radionuclides (40 CFR 141.16). While some of these standards apply to both community water systems and noncommunity water systems, the standards for radionuclides in items (5) and (6) are applicable only to community water systems.

As an example of the maximum contaminant levels prescribed by the EPA, a summary of the standards for a few selected contaminants is given in Table 4. A more complete list of contaminants regulated by EPA in 40 CFR 141 can be found in Chapter 6, Section III.

V. SUMMARY

This chapter has presented a brief description of the methods by which the goals of environmental statutes are translated to technical and measurable regulations and standards which can be implemented. The procedures and methods of using these standards in the NEPA process are presented in Chapter 5 (How to Write Environmental Documents). The presentation of standards and regulations has been necessarily brief because the complete regulations and standards promulgated to achieve the statutory goals fill literally several thousand pages and are beyond the scope of this book. In fact, this chapter is quite brief because it excludes the discussion of standards from NCA, RCRA, and CERCLA which are very important for projects that involve noise impacts, hazardous waste handling, and abandoned poisonous waste sites. However, all the applicable regulations and standards can be found in Code of Federal Regulations purchasable at any U.S. Government Printing Office (U.S. GPO) bookstore outlets located in major cities of the U.S.

The air quality and water quality models used to estimate pollutant concentrations will also fill several volumes and are beyond the scope of this book. The qualified environmental professional would readily know which model to use and when it was approved by the EPA.

The literature on control technology available to meet the emission regulations literally fills millions of pages, and such detail is beyond the scope of this book. A brief introduction to the process of selecting control technologies is presented later in Chapter 6.

* Groundwater quality is regulated by the states in which the groundwater is impacted and by the EPA by other federal regulations. When groundwater is used as a source of community water supply, 40 CFR 141 applies to the groundwater supplied to the community at the faucet.

Table 4
PARTIAL LISTING OF MAXIMUM CONTAMINANT LEVELS ACCORDING TO FEDERAL DRINKING WATER STANDARDS (40 CFR 141)

Contaminant name		Maximum contaminant level	
		mg/ℓ	pCi/ℓ
Inorganic chemical[a]			
Arsenic		0.05	
Barium		1.00	
Cadmium		0.01	
Chromium		0.05	
Lead		0.05	
Mercury		0.002	
Nitrate as N		10.0	
Nitrate as NO_3^-		0.45	
Selenium		0.01	
Silver		0.05	
Radionuclides[b]			
Tritium	(^{3}H)		20,000
Carbon-14	(^{14}C)		1,000
Cobalt-60	(^{60}Co)		100
Strontium-90	(^{90}Sr)		8
Technecium-99	(^{99}Tc)		900
Ruthenium-106	(^{106}Ru)		30
Iodine-129	(^{129}I)		0.6
Cesium-137	(^{137}Cs)		60
Uranium-238	(^{238}U)		100
Plutonium-239/240	($^{239/240}Pu$)		7

[a] Applicable to both community and noncommunity water systems.

[b] Applicable only to community water systems at the faucet (calculated using the method specified in EPA Regulation 40 CFR 141). If two or more radionuclides are present, the radiation dose to the whole body or to any organ resulting from the sum of their radioactivity shall not exceed 4 mrem/year.

Chapter 4

THE NEPA PROCESS

I. INTRODUCTION

This chapter describes four major elements that need to be understood in implementing NEPA: (1) the environmental documents and what they are, (2) the NEPA process itself, (3) federal and state permits needed to supplement environmental documents, and (4) role of nonfederal agencies in the NEPA process. NEPA mandates that decisions that potentially impact the health and welfare of the people shall be made with full awareness of the pros and cons of the outcome of the decision. The procedures involved in understanding the pros and cons are defined in the CEQ regulations (40 CFR 1500 to 1508). The process of applying these regulations in a systematic manner to the decision-making procedures is called the NEPA process. The discussion of the process begins below with the first step, i.e., an understanding of the environmental documents, supporting documents, and who prepares which document.

II. WHAT ARE THE ENVIRONMENTAL DOCUMENTS?

The CEQ regulations recognize four distinct NEPA documents as ''environmental documents'' in 40 CFR 1508.10. These documents are

1. Environmental assessment (EA)
2. Finding of no significant impact (FONSI)
3. Notice of intent (NOI)
4. Environmental impact statement (EIS)

By law, these four documents are subject to public scrutiny. These are also subject to judicial review, although judicial review of the notice of intent perhaps would not be worthwhile.

Aside from these CEQ-designated environmental documents, it is important to recognize that the environmental documentation process is the method available for complying with NEPA, the environmental protection law of the land. According to NEPA and CEQ, all federal agencies have to perform environmental compliance planning and documentation for a proposed project. The compliance process should begin at the project level.

The project can be any one of a variety of activities, and the type of project varies from agency to agency. For example, if the U.S. EPA wishes to promulgate major new regulations, it has to evaluate alternative methods of attaining the goals that the proposed regulations would achieve and evaluate environmental and socioeconomic impacts of the proposed and any alternative regulations. This process, depending on the range of impact of the proposed regulations, could require the EPA to comply fully with the NEPA process including preparation of an environmental impact statement and public hearings as part of public participation requirements. Another example could be that the U.S. Department of Energy wishes to build a new reactor to produce plutonium. Since building a new nuclear reactor is quite a major issue in the public mind, full NEPA process would have to be followed. In both these examples, the federal agencies involved are the lead agencies and these agencies also are proposing and initiating major activities.

There are instances in which federal agencies are not initiators of major proposals, rather they license the construction and operation of major proposals on the basis of the application

Table 1
WHO PREPARES WHICH DOCUMENTS?

Industrial applicant requesting license	Lead agencies of government (licensing authority)
Environmental reports	Environmental assessments
Safety Analysis reports	Notice of intent
Environmental evaluations	FONSI
Alternatives analysis	EIS
Socioeconomic analysis	ROD
Technology alternatives report	

for licensing they receive from private industry or public utilities. In such cases, these federal agencies also would have to comply with the NEPA process. To respond to these varying needs, each federal agency has developed and published procedures from implementation of the requirements of NEPA.

In addition to the four environmental documents that lead agencies prepare, there are other supporting documents that an industrial applicant requesting a license from a regulatory agency would need to prepare. Table 1 provides examples of the types of documents that different organizations may need to prepare for a proposed project.

III. WHAT IS THE NEPA PROCESS?

The NEPA process begins with the preparation of a plan for environmental compliance at the project level by the lead agency or its EIS contractor. Such a plan would be designed to encompass all relevant federal and state environmental requirements including NEPA and other statutes such as the Clean Air Act, Clean Water Act, and the Endangered Species Act. Plans should be developed early as an integral part of the total planning effort. For a given action, a number of environmental requirements may be applicable. The compliance plan should allow for and discuss how the complex trade-offs of time, program risk, financial and environmental costs, quality of environmental review, and other considerations are to be resolved. Failure to properly comply with these environmental regulations and statutes invites lawsuits, delays, loss of resources, and environmental degradation. The plan should indicate how proper compliance with the regulations and statutes will be achieved.

Compliance at the project level is the responsibility of the lead agency and is illustrated in Figure 1. NEPA compliance, as shown in Figure 1, begins with the publication of the notice of intent (NOI) to prepare an EIS. For projects that originate in a federal agency, there are usually several predecisional steps before an NOI is published. These predecisional steps include considerations such as (1) whether the project is large and complex enough to require legislative authorization by the U.S. Congress, (2) whether the impacts of the project are such that an EIS is not required, and (3) whether the action is listed in those NEPA procedures of that agency as one that normally requires an EIS. The key question should be: Is it a major federal action significantly affecting the quality of the human environment?

A. Legislative EIS (LEIS)

When a project is complex or enormous and ''politically visible'' and requires legislative authorization because the agency is recommending legislation, the resulting legislative proposal itself needs an environmental impact statement as specified in Section 102 of the NEPA, and elaborated upon in 40 CFR 1506.9 (Proposals for Legislation). At the option of the lead agency, the overall efforts to prepare an LEIS could be less than the efforts needed for a full scale EIS because

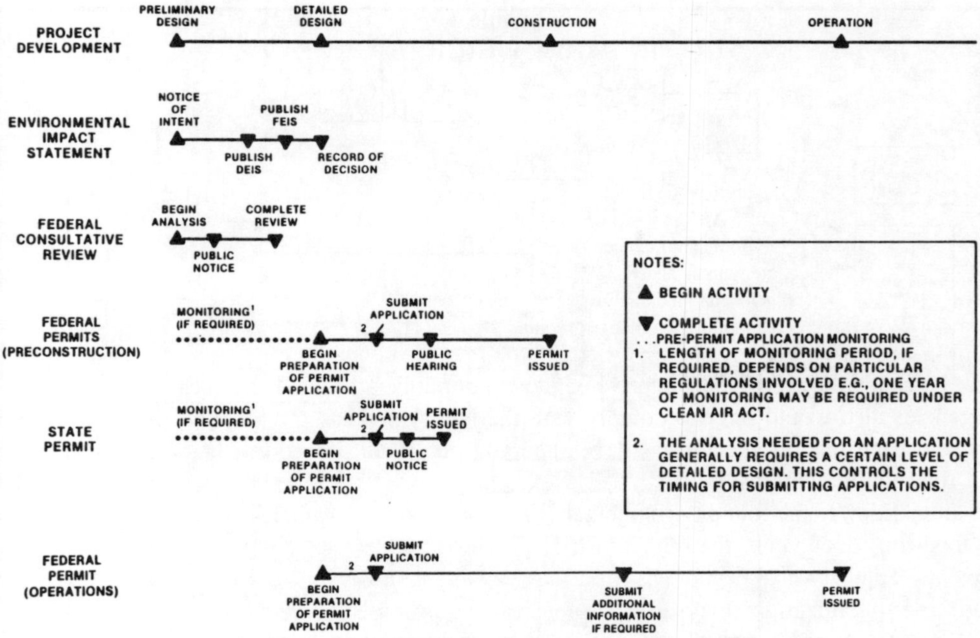

FIGURE 1. Project level compliance — phased compliance.[33]

1. LEIS need not involve a scoping process.
2. LEIS does not require a Notice of Intent.
3. Draft EIS may be adequate; a final EIS may not be required.

The LEIS (as a draft EIS) must be submitted to Congress within 30 days after submission of the submission of a proposal for legislation. Some lead agencies may opt to prepare the LEIS with a scoping process, NOI, and a final EIS.

As defined by CEQ regulations (40 CFR 1508.17), ''Legislation'' includes a bill or legislative proposal to Congress developed by or with the significant cooperation and support of a Federal agency, but does not include requests for appropriations. The test for significant cooperation is whether the proposal is in fact predominantly that of the agency rather than another source.

Although very few legislative EISs have been filed by federal agencies, court decisions indicate that successful challenges on the basis of NEPA requirements can be made to a legislative initiative that is ''predominantly'' that of a federal agency and there is an absence of a NEPA strategy that integrates the various levels of environmental review. When a federal agency is not ''predominantly'' involved in a legislative proposal (i.e., when the legislative initiative is that of Congress) or the agency is only requesting an appropriation to undertake an action under existing law, only a normal EIS (i.e., project- or programmatic-level EIS) may be required, but not a legislative EIS.

There have not been many LEISs published. In general, the LEIS requirement appears to be more honored in the breach than in the observance. On the other hand, the normal EIS requirement has been more or less religiously followed by federal agencies. Unlike the LEIS, the full EIS process (for a project or a program) requires

1. Notice of intent
2. Scoping meetings (as necessary)
3. Both DEIS and FEIS publications after public comment process
4. Publication of the record of decision

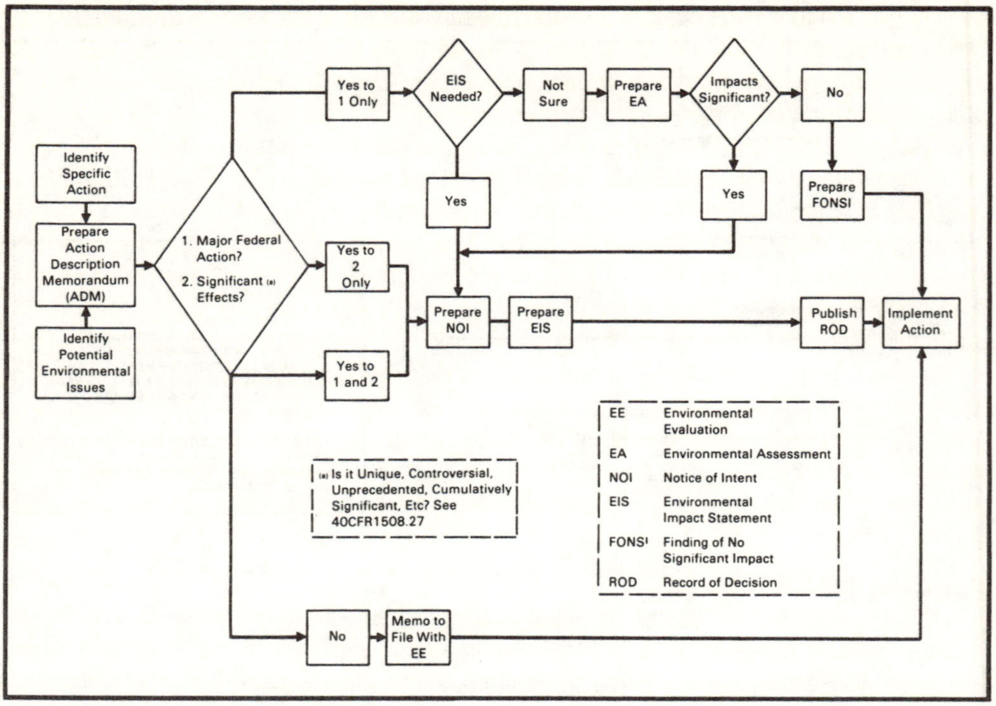

FIGURE 2. Decision steps in NEPA compliance for any size project not categorically excluded from an EA/EIS requirement.

The need for scoping meetings is a function of the prevailing and potential controversy, the complexity, and the precedents on the proposed action. The lead agency has the discretion to decide on the need for meeting based on these factors.

There are at least four ways that a project can take birth:

1. By congressional legislation in which the Congress initiates a project (this will not require a LEIS since the Congress initiated it)
2. By legislative authorization of a proposal sent to the Congress by a federal agency (this requires LEIS)
3. By agency authorization of a project from available funds within a federal agency (this usually requires EA/EIS)
4. By a federal agency licensing and permitting the construction and operation of a project in response to license application from a private industry, such as, for example, petroleum refiners, or a public utility wishing the construction of a coal or nuclear power plant (this requires either an EA or an EIS)

B. Full-Scale EIS (Normal EIS)

Once a project becomes real (in the sense that commitment of funds and resources are provisionally made), a decision maker has to decide to "implement the action". By law, the decision maker cannot decide and implement the action without satisfying the requirements of NEPA. Thus, the wheels of NEPA will be set in motion in accordance with the decision steps shown in Figure 2.

As shown in Figure 2, the decision to implement the action need not always and does not automatically require an EIS. Unless the action is unique, controversial, unprecedented, or cumulatively significant, as described in 40 CFR 1508.27 (see Appendix B), and/or unless the action is a major federal action, as described in 40 CFR 1508.18 (see Appendix B), an EIS is not normally required. This part of the decision analysis is very important to the progress of the project because a wrong analysis can lead to significant and costly delays in the implementation of the project. The wrong analysis can take either form: (1) a decision to prepare an EIS when it is not required, and (2) a decision not to prepare an EIS when one is actually required. **Therefore, a legal review of the decision process and results is essential.** It requires about 10 to 40 months of effort to prepare an EIS depending on the complexity of the project, availability of the data, and the degree of public controversy associated with the project. Also, in dollar terms, the EIS can cost anywhere from $200,000 to $4,000,000. If a decision is made not to write an EIS when one is actually required, the public can take (and have taken) the decision maker to court after implementation of the project has begun and force the decision maker to write the EIS and publish a record of decision. This process could cost (and has cost) projects several times the duration and dollars required to complete the EIS. Therefore, abiding by the decision analysis method described in Figure 2 is very critical to the progress of the project. More details of the application of the decision analysis are presented in Chapter 5. (How to Write Environmental Documents).

C. Tiering in NEPA Process

In the decision analysis step at this stage of a project, another important time- and dollar-saving procedure available to the lead agency is called "tiering". Tiering may be defined as a method of reducing paperwork by judicious use of already published EISs of broader scope to the best advantage of the NEPA process. This can be done by incorporating the published EIS by reference in the subsequent specific action EIS. One example of tiering is to prepare an EIS supplement to an existing EIS, thus avoiding, at least, a lengthy description of the affected environment portion of the EIS. The CEQ defines tiering in 40 CFR 1508.28. CEQ allows tiering as an appropriate method of NEPA compliance when the sequence of environmental impact statements or environmental analysis is as follows:

(a) From a program, plan, or policy environmental impact statement to a program, plan, or policy statement or analysis of lesser scope or to a site-specific statement or analysis.

(b) From an environmental impact statement on a specific action at an early stage (such as need and site selection) to a supplement (which is preferred) or a subsequent statement or analysis at a later stage (such as environmental mitigation). Tiering in such cases is appropriate when it helps the lead agency to focus on the issues which are ripe for decision and exclude from consideration issues already decided or not yet ripe.

The analytical process to be used in determining the need for tiering is shown in Figure 3. Before even examining the need for tiering, the action may be one that qualifies for "categorical exclusion", as defined in 40 CFR 1508.4, and should be investigated.

IV. DETAILED DESCRIPTION OF ENVIRONMENTAL DOCUMENTS

In the preceding discussions, the four documents designated by the CEQ as environmental documents were identified as EA, FONSI, NOI, and EIS. Another document that is published as a public document by the lead agency is the record of decision (ROD). These five documents are described below as to their purpose and content.

A. Environmental Assessment

1. Purpose

The Council on Environmental Quality (CEQ) regulations 40 CFR 1508.9 state that environmental assessment:

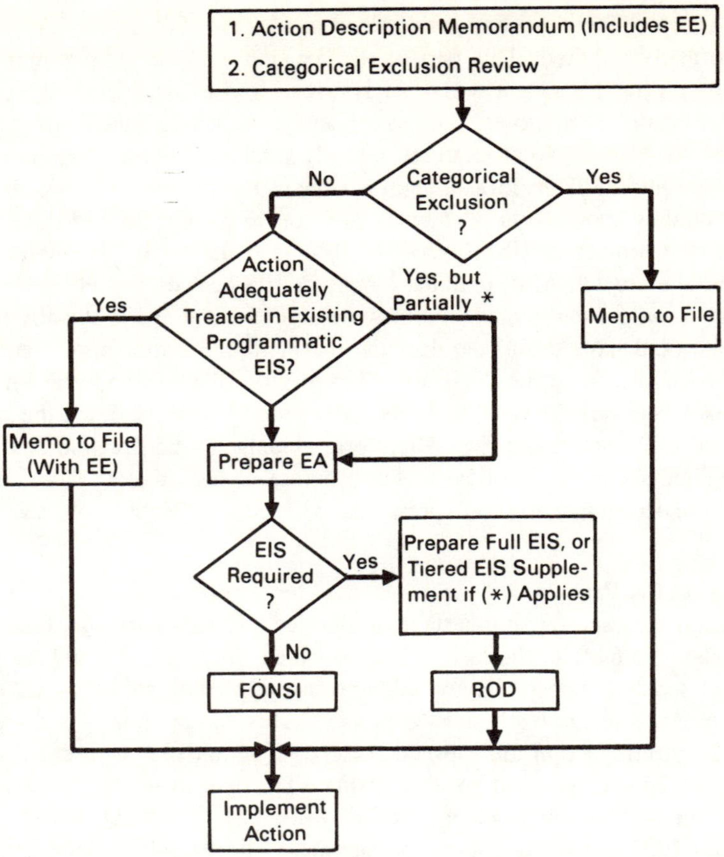

FIGURE 3. Tiering to existing documents

"(a) Means a concise public document for which a Federal agency is responsible that serves to:

(1) Briefly provide sufficient evidence and analysis for determining whether to prepare an environmental impact statement or a finding of no significant impact.

(2) Aid an agency's compliance with the Act when no environmental impact statement is necessary.

(3) Facilitate preparation of a statement when one is necessary.

(4) Shall include brief discussions of the need for the proposal, of alternatives as required by section 102(2)(E), of the environmental impacts of the proposed action and alternatives, and a listing of agencies and persons consulted."

2. Format and Content

An environmental assessment (EA) shall, as appropriate, contain the following information: a clear and concise description of the proposed action, including drawings, maps, and charts, if directly pertinent to analyzing the environmental consequences of the proposed action; a description of the existing environment affected by the proposed action only in sufficient detail to permit a meaningful evaluation of the potential environmental consequences of the proposed action; an assessment of the probable impacts of the proposed action including direct and indirect effects and those adverse impacts which cannot be avoided should the proposal be implemented; an evaluation of the probable cumulative and long-term environmental effects including any beneficial impacts; an assessment of the risk of credible accidents; a discussion of the relationship of the proposed action to any applicable federal, state, regional, or local land use plans and policies likely to be affected; and a brief description of reasonable alternatives to the proposed action and their probable environmental effects.

B. Finding of No Significant Impact

1. Purpose and Content

The CEQ regulations 40 CFR 1508.13 state that: a finding of no significant impact (FONSI) is a document by a federal agency briefly presenting the reasons why an action, not otherwise excluded (1508.4), will not have a significant effect on the human environment and for which an environmental impact statement will therefore not be prepared. It shall include the environmental assessment or a summary of it and shall note any other environmental documents related to it (1501.7(a)(5)). If the assessment is included, the finding need not repeat any of the discussion in the assessment, but may incorporate it by reference.

2. General Outline

1. Declaration of finding (including reference to the EA and other applicable documents)
2. Description of the proposed action
3. Alternatives considered
4. Description of impacts and justification for conclusion of no significance
5. Availability of EA
6. Contact

3. Circulation and Dissemination

The Council on Environmental Quality (CEQ) regulations 40 CFR 1501.4(e) state:

"(1) The agency shall make the finding of no significant impact available to the affected public as specified in § 1506.6.

(2) In certain limited circumstances, which the agency may cover in its procedures under § 1507.3, the agency shall make the finding of no significant impact available for public review (including State and areawide clearinghouses) for 30 days before the agency makes its final determination whether to prepare an environmental impact statement and before the action may begin. The circumstances are:

(i) The proposed action is, or is closely similar to, one which normally requires the preparation of an environmental impact statement under the procedures adopted by the agency pursuant to § 1507.3, or

(ii) The nature of the proposed action is one without precedent."

Diligent effort shall be made to make the FONSI available to the public in the following manner consistent with the Council on Environmental Quality (CEQ) regulations 40 CFR 1506.6. The FONSI will be

1. Published in the *Federal Register* in the case of an action of national concern.
2. Mailed to those who requested it on an individual action.
3. Mailed to national organizations reasonably expected to be interested in the manner.

Where the action is primarily of local concern, the notice will also be

1. Sent to state- and area-wide clearinghouses pursuant to Office of Management and Budget Circular A-95 (revised).

2. Published in local circulation newspapers.
3. Sent to potentially interested community organizations.
4. Distributed through other local media.
5. Published in newsletters that may be expected to reach potentially interested persons.
6. Mailed directly to owners and occupants of nearby or affected property.
7. Posted on and offsite in the area where the action is to be located.
8. Sent to Indian tribes when effects may occur on reservations.
9. Distributed in accordance with the public notice procedures of the affected states for comparable actions.

As an example, an actual FONSI published by the U.S. Postal Service in *The Washington Post,* Wednesday, May 14, 1984, is shown in Figure 4.

C. Notice of Intent

The NOI is the mechanism (employed by the lead agency) that notifies the public of the intent to prepare an EIS and requests the public to participate in the formulation of the issues to be addressed in the EIS.

As soon as practicable after the decision to prepare an environmental impact statement (EIS), and before the scoping process, a notice of intent shall be published in the *Federal Register* unless there is a lengthy period of time between the time of decision and preparation of the draft statement. Where such a delay may exist, the notice of intent may be published at a reasonable time in advance of preparation. Diligent effort shall be made to notify and involve the public in the following manner consistent with the Council on Environmental Quality (CEQ) regulations 40 CFR 1506.6.

1. Recommended General Outline

1. Agency identification
2. Proposed action
3. Summary of action and purpose of notice (include address for submission of comments and/or contact for additional information)
4. Dates and location of any hearings
5. Background information
   a. Site description
   b. Proposed action including need
   c. Identification of significant environmental (and socioeconomic) issues
   d. Timing
6. Alternatives including the proposed action
7. Comments and hearings
   a. Invitations
   b. Procedures
   c. Dates, times, location
   d. Availability of related documents

2. Distribution

As a minimum, the notice will be

1. Published in the *Federal Register*
2. Mailed to those who requested it on an individual action.
3. Mailed to national organizations reasonably expected to be interested in the matter.

UNITED STATES POSTAL SERVICE
475 L'Enfant Plaza, SW
Washington, DC 20260–6236

FINDING OF NO SIGNIFICANT IMPACT

ACTION:

The United States Postal Service is planning to build a new General Mail Facility (GMF) for Southern Maryland in Largo, Maryland, adjacent to the existing Bulk Mail Center (BMC) and Vehicle Maintenance Facility. The new GMF will replace the existing leased Prince George's Management Sectional Center (MSC) in Riverdale and the Prince George's Interim Facility, which is located on the opposite side of Edgeworth Drive and Hampton Park Boulevard from the BMC. Both existing facilities are deficient in workroom, office and support space, and have substandard working conditions. Following the completion of the new facility, the existing Prince George's MSC will be vacated and the old building subleased; the leased Interim Facility also will be vacated and the lease terminated.

The project consists of construction of a new GMF/MSC building containing approximately 250,000 gross square feet of floor space which will be located on Edgeworth Drive approximately .72 miles south of Central Avenue (MD. Rt. 214) and adjacent to the western right-of-way line of the Capital Beltway. The construction of the new facility will require an additional 8.66 acres of land west of and contiguous to the existing 83-acre BMC property. The property has access from Edgeworth Drive and the future extension of Hampton Park boulevard and is surrounded primarily by industrially zoned land.

FINDING:

This finding is based on the Environmental Assessment, dated July 1982, that was prepared for the proposed project. The project was found to have no significant impacts on either the physical or the cultural environment of the project area. Where adverse impacts were found, they were determined to be minor and will be mitigated by proper design and construction practices.

Therefore, I hereby find and declare that this action will not have a significant effect on the human environment and that it will not require the preparation of an Environmental Impact Statement.

Frank P. Rowan
Director
Office of Program Planning
Real Estate and Buildings Department
Telephone: 202/245-4304

This Finding of No Significant Impact was issued to various Government bodies in December 1983; however, its publication in the newspaper was inadvertently delayed until now.

FIGURE 4. FONSI example. (From *The Washington Post,* Wednesday, May 14, 1984.)

Where the action is primarily of local concern, the notice will also be

1. Sent to state- and area-wide clearinghouses pursuant to Office of Management and Budget Circular A-95 (revised).
2. Published in local circulation newspapers.
3. Sent to potentially interested community organizations.

4. Distributed through other local media.
5. Published in newsletters that may be expected to reach potentially interested persons.
6. Mailed directly to owners and occupants of nearby or affected property.
7. Posted on and offsite in the area where the action is to be located.

D. Environmental Impact Statement

The lead agency will prepare EISs consistent with the following Council on Environmental Quality (CEQ) regulations 40 CFR 1502.9 through 1502.18:

1. Environmental impact statements shall be analytic rather than encyclopedic. ''Analytic'' means factual information supported by analysis and data, rather than qualitative and verbose statements.
2. Impacts shall be discussed in proportion to their significance. There shall be only brief discussion of other than significant issues. As in a finding of no significant impact, there should be only enough discussion to show why more study is not warranted.
3. Environmental impact statements shall be kept concise and shall be no longer than absolutely necessary to comply with NEPA and with the CEQ NEPA Regulations 40 CFR 1500 to 1508. Length should vary first with potential environmental problems and then with project size.
4. Environmental impact statements shall state how alternatives considered in it and decisions based on it will or will not achieve the requirements of Sections 101 and 102(1) of the act and other environmental laws and policies.
5. The range of alternatives discussed in environmental impact statement shall encompass all the alternatives to be considered by the ultimate lead agency decision maker.
6. The lead agency shall not commit resources prejudicing selection of alternatives before making a final decision.
7. Environmental impact statements shall serve as the means of assessing the environmental impact of proposed lead agency actions, rather than justifying decisions already made.

1. Format

The lead agency shall use a format for environmental impact statements which will encourage good analysis and clear presentation of the alternatives, including the proposed action. The following standard format for environmental impact statements should be followed unless there is a compelling reason to do otherwise:

(a) Cover sheet
(b) Summary
(c) Table of contents
(d) Purpose of and need for action
(e) Alternatives including proposed action (Sections 102(2)(C)(iii) and 102(2)(E) of the act)
(f) Affected environment
(g) Environmental consequences (especially Sections 102(2)(C)(i), (ii), (iv), and (v) of the act)
(h) List of preparers
(i) List of agencies, organizations, and persons to whom copies of the statement are sent
(j) Index
(k) Appendixes (if any)

If a different format is used, it shall include paragraphs (a), (b), (c), (h), and (i), and (j) of this section and shall include the substance of paragraphs (d), (e), (f), (g), and (k) of this section, as further described in 40 CFR 1502.11 to 1502.18, in any appropriate format.

2. Section 1502.11, Cover Sheet

The cover sheet shall not exceed one page. It shall include the following:

(a) A list of the responsible agencies, including the lead agency and any cooperating agencies.
(b) The title of the proposed action that is the subject of the statement (and, if appropriate, the titles of related cooperating agency actions), together with the state(s) and county(ies) (or other jurisdiction if applicable) where the action is located.
(c) The name, address, and telephone number of the person at the lead agency who can supply further information.
(d) A designation of the statement as a draft, final, or draft or final supplement.
(e) A one-paragraph abstract of EIS.
(f) The date by which comments must be received (computed in cooperation with EPA under 40 CFR 1506.10).

3. Section 1502.12, Summary

Each environmental impact statement shall contain a summary which adequately and accurately summarizes the statement. The summary shall stress the major conclusions, areas of controversy (including issues raised by agencies and the public), and the issues to be resolved (including the choice among alternatives). The summary will normally not exceed 15 pages.

4. Section 1502.13, Purpose and Need

The statement shall briefly specify the underlying purpose and need to which the lead agency is responding in proposing the alternatives, including the proposed action.

5. Section 1502.14, Alternatives Including Proposed Action

This section is the heart of the environmental impact statement. Based on the information and analysis presented in the sections on the affected environment (40 CFR 1502.15) and the environmental consequences (40 CFR 1506.15), it should present the environmental impacts of the proposal and the alternatives in comparative form, thus sharply defining the issues and providing a clear basis for choice among options by the lead agency and the public. In this section, the lead agency shall:

(a) Rigorously explore and objectively evaluate all reasonable alternatives and, for alternatives which were eliminated from detailed study, briefly discuss the reasons for their having been eliminated.
(b) Devote substantial treatment to each alernative considered in detail, including the proposed action, so that reviewers may evaluate their comparative merits.
(c) Include reasonable alternatives not within the jurisdiction of (d) and (e).
(d) Include the alternative of no action.
(e) Identify the preferred alternative or alternatives, if one or more exist, of the lead agency in the draft statement and identify such alternative in the final statement unless another law prohibits the expression of such a preference.
(f) Include appropriate mitigation measures not already included in the proposed action or alternatives.

The preferred alternative of the lead agency may be defined as the alternative which the lead agency has selected for adoption. According to the CEQ regulations, the lead agency may *always* choose a preferred alternative based on *any* considerations. The decision maker is NEVER required to select the environmentally preferable alternative. Identification of the environmentally preferable is required in the record of decision, but that can be quite different from the preferred alternative. However, when all alternatives have the same environmental advantages/disadvantages, the lead agency may select from among those alternatives a preferred alternative based on schedule, technology, cost, and policy considerations.

If cost/benefit considerations are to be used as part of a lead agency decision, the EIS should identify those considerations and indicate how they were relevant and important to a decision (see CEQ 40 CFR 1502.23). It is important, however, that cost/benefit considerations should not be the sole criterion for selecting a preferred alternative; as a rule, environmental impacts should have overriding priority in the decision to select an alternative for final implementation. A rare exception to this rule would be that the cost/benefit ratio is unreasonably high (for example, say, greater than $10 million to reduce the risk of one potential cancer, mutation, or death). The definition of "unreasonably high" varies with the times.

6. Section 1502.15, Affected Environment

The environmental impact statement shall succinctly describe the environment of the area(s) to be affected or created by the alternatives under consideration. The descriptions shall be no longer than is necessary to understand the effects of the alternatives. Data and analyses in a statement shall be commensurate with the importance of the impact, with less important material summarized, consolidated, or simply referenced. The lead agency shall avoid useless bulk in statements and shall concentrate effort and attention on important issues. Verbose descriptions of the affected environment are themselves no measure of the adequacy of an environmental impact statement.

7. Section 1502.16, Environmental Consequences

This section forms the scientific and analytic basis for the comparisons under 40 CFR 1502.14. It shall consolidate the discussions of those elements required by Sections 102(2)(c)(i), (ii), (iv), and (v) of NEPA which are within the scope of the statement and as much of Section 102(2)(C)(iii) as is necessary to support the comparisons. The discussion will include the environmental impacts of the alternatives, including the proposed action; any adverse environmental effects which cannot be avoided should the proposal be implemented; the relationship between short-term uses of man's environment and the maintenance and enhancement of long-term productivity; and any irreversible or irretrievable commitments of resources which would be involved in the proposal should it be implemented. This section should not duplicate discussions of 40 CFR 1502.14. It shall include discussions of the following:

(a) Direct effects and their significance (40 CFR 1508.8).
(b) Indirect effects and their significance (40 CFR 1508.5).
(c) Possible conflicts between proposed action and the objectives of federal, regional, state, and local (and, in the case of reservation, Indian tribe) land use plans, policies, and controls for the area concerned (40 CFR 1506.2(d)).
(d) The environmental effects of alternatives, including the proposed action (the comparison under 40 CFR 1502.14 will be based on this discussion).
(e) Energy requirements and conservation potential of various alternatives and mitigation measures.
(f) Natural or depletable resource requirements and conservation potential of various alternatives and mitigation measures.

(g) Urban quality, historic and cultural resources, and the design of the built environment, including the reuse and conservation potential of various alternatives and mitigation measures.
(h) Means to mitigate adverse environmental impacts (if not fully covered under 40 CFR 1502.14(f)).

8. Section 1502.17, List of Preparers

The environmental impact statement shall list the names of persons, together with their qualifications (expertise, experience, professional disciplines), who were primarily responsible for preparing the environmental impact statement or significant background papers, including basic components of the statement (40 CFR 1502.6 and 40 CFR 1502.8). Where possible, the persons responsible for a particular analysis, including analyses in background papers, shall be identified. Normally the list will not exceed two pages.

9. Section 1502.18, Appendixes

If the lead agency prepares an appendix to an environmental impact statement, the appendix shall

(a) Consist of material prepared in connection with an environmental impact statement (as distinct from material which is not so prepared and which is incorporated by reference (40 CFR 1502.21)).
(b) Normally consist of material which substantiates any analysis fundamental to the impact statement.
(c) Normally be analytic and relevant to the decision to be made.
(d) Be circulated with the environmental impact statement or be readily available on request.

The appendix need not be published as a separate volume, but can be incorporated into one volume. If the appendix is too large, and if it increases the size of the EIS, a separate volume for the appendix is the best way to go.

10. References for the EIS

Although the CEQ recommended format does not include a listing of cited references, references are an important part of the EIS and must be cited at the end of the EIS. The EIS is a technical-scientific-political document written for policy decisions and public consumption. The quality of scientific analysis is dependent on the quality of background data used as evidenced by the references employed. Also, if an EIS is tiered, the references will have to cite the main EIS to which the tiered EIS is a supplement.

11. Index

The CEQ regulations (40 CFR 1502.10) specify the index as a part of the recommended format. Index is a tool very useful to the public in looking up issues that concern them and finding a quick response. The use of an index is strongly recommended.

In closing the discussion on the format and content of the EIS, it is well to remember that exercising too much creativity on devising new formats is, in general, futile. The format recommended by CEQ has served well for nearly 20 years. Therefore, it is wise to use the CEQ format and save the creative talents of the EIS team to prepare an excellent technical analysis of the main issues identified through the NOI process. It is also important the EIS be written in plain language, using appropriate graphics, so that the decision makers and the public can readily understand them.

12. Circulation of Environmental Impact Statements

An integral and important part of the NEPA process involving environmental documentation is the circulation of EIS. It is not enough if the lead agency prepares the EIS, it is critical that the agency take steps to circulate the EIS, receive comments, and respond to the comments. This active communication step on the part of the bureaucracy is a novel aspect of the modern industrial era wherein the government admits that its decision process can benefit by interaction with the public and other agencies of the government.

The Council on Environmental Quality (CEQ) regulations 40 CFR 1502.19 state the following regarding circulation of EISs:

''Agencies shall circulate the entire draft and final environmental impact statements except for certain appendices as provided in § 1502.18(d) and unchanged statements as provided in § 1503.4(c). However, if the statement is unusually long, the agency may circulate the summary instead, except that the entire statement shall be furnished to:

(a) Any Federal agency which has jurisdiction by law or special expertise with respect to any environmental impact involved and any appropriate Federal, State, or local agency authorized to develop and enforce environmental standards.
(b) The applicant, if any.
(c) Any person, organization, or agency requesting the entire environmental impact statement.
(d) In the case of a final environmental impact statement, any person, organization, or agency which submitted substantive comments on the draft.

If the agency circulates the summary and thereafter receives a timely request for the entire statement and for additional time to comment, the time for that requestor only shall be extended by at least 15 days beyond the minimum period.''

E. Record of Decision (ROD)

At the conclusion of the NEPA process in which an EIS has been prepared, the lead agency is required to prepare a concise public record of decision (40 CFR 1505.2). The ROD, which must be published in the *Federal Register,* will

(a) State what the decision is.
(b) Identify all alternatives considered by the lead agency in reaching its decision, specifying the alternative or alternatives which were considered to be environmentally preferable. The lead agency may discuss preferences among alternatives based on relevant factors, including economic and technical considerations and its statutory missions. The agency shall identify and discuss all such factors, including any essential considerations of national policy which were balanced by it in making its decision, and state how those considerations entered into its decision.
(c) State whether all practicable means to avoid or minimize environmental harm from the alternative selected have been adopted and, if not, why they are not. A monitoring and enforcement program shall be adopted and summarized where applicable for any mitigation.

V. OTHER PERMITTING NEEDS

A discussion of the NEPA process would be incomplete without addressing the permits required to construct and operate projects, identified earlier in Figure 1. Four types of permits need to be considered:

- Federal permits before construction can begin
- State permits before construction begins

Table 2
SCREENING FOR THE APPLICABILITY OF OTHER FEDERAL AND STATE ENVIRONMENTAL REGULATIONS AND PERMITS

Question	If yes, examine need for following permit, consultation, or review
1. Does the project involve emissions of pollutants to the air?	PSD permit, NSPS permit possible
2. Does the action involve the discharge of pollutants into a waterway of the U.S. or the ocean?	NPDES permit
3. Will the project require the discharge of dredged or fill materials into waters of the U.S.?	Corps of Engineers permit and state water quality certification
4. Does the project require structures (piers, power lines, etc.) or work such as dredging in navigable waters of the U.S.?	Corps of Engineers permit
5. Is the action located with a floodplain or in a wetland area?	Floodplains/wetlands review
6. Might the action impact a coastal zone area?	State Coastal Zone Management consultations
7. Are there any threatened or endangered species or their critical habitats located in the area of the proposed action?	Endangered species consultation
8. Will the project modify or impact a waterway?	Fish and wildlife review
9. Will the action affect a wild and scenic river area or areas designated as potential additions to the Wild and Scenic Rivers System?	Review by SEC/DOI or SEC/USDA
10. Are there historic sites in the project area?	Consultation with state HPO and ACHP
11. Might the project have an effect on a water resource?	Water Resources Council review
12. Does the project require the handling, disposal, or transport of hazardous materials?	RCRA permit
13. Does the project involve the injection of a substance into the ground?	UIC permit

From Environmental Compliance Guide, Vol. 1 and 2, DOE/EV-0132, U.S. Department of Energy, Washington, D.C., 1981.

- State permits for operations
- Federal permits for operations

The permitting requirements are dictated by a number of federal and state environmental, safety, and health statutes. Examples are

- Clean Water Act — federal and state laws require permits including an NPDES permit.
- Clean Air Act — permits for radioactive and nonradioactive discharges under both federal and state law. The Prevention of Significant Deterioration (PSD) permit falls under this act.
- Solid Waste Disposal Act as amended by Resource Conservation and Recovery Act (RCRA) — permits may be required under both federal and state law.
- Noise Control Act — permits may be required under both federal and state law.

Examples of other permits that may be required as a matter of comity are (1) a permit to appropriate water under state law; (2) underground injection control permit; and (3) permit for leaking underground storage tanks (LUST). A preliminary screening of permit needs can be aided by reference to Table 2 which lists the type of questions that a permitting agency would ask and the type of possible permits needed. Not all of the permits would apply to a particular project. Hence, the question-and-answer format of Table 2 would also assist in eliminating unnecessary work on unneeded permits.

Table 3
ENVIRONMENTAL IMPACT STATEMENTS[a] FILED BY FEDERAL AGENCIES DURING 1978 TO 1984

Agency	Calendar year						
	1978	1979	1980	1981	1982	1983	1984
U.S. Department of Agriculture	225	172	104	102	89	59	65
U.S. Department of Commerce	56	54	53	36	25	14	24
U.S. Department of Defense	—	1	1	1	1	1	0
Air Force	5	8	3	7	4	6	5
Army	16	40	9	14	3	6	5
U.S. Army Corps of Engineers	182	182	150	186	127	119	116
Navy	6	11	9	10	6	4	9
U.S. Department of Energy	43	28	45	21	24	19	14
Environmental Protection Agency	77	84	71	96	63	67	42
General Services Administration	27	13	11	13	8	1	0
U.S. Department of Housing and Urban Development	263	170	140	140	93	42	13
U.S. Department of the Interior	91	126	131	107	127	146	115
Interstate Commerce Commission	9	8	5	2	2	2	1
Nuclear Regulatory Commission	25	19	17	47	31	11	11
Regional River Basin Commissions	6	12	8	8	2	0	0
Tennessee Valley Authority	6	9	6	4	—	2	1
U.S. Department of Transporation	288	277	189	221	183	169	147
Other federal agencies, commissions, councils	30	59	14	18	20	9	9
Total, all federal agencies	1,355	1,273	966	1,033	808	677	577

[a] Number of environmental impact statements including draft EIS, EIS supplements, and final EISs filed during specific year. Some proposed projects may have several drafts and final EIS filed over a period of years.

From Environmental Quality, 15th Annu. Rep. Council on Environmental Quality, Council on Environmental Quality, Washington, D.C., 1984.

VI. ENVIRONMENTAL DOCUMENTATION AT NONFEDERAL AGENCIES

In the preceding discussions, only the lead agency has been mentioned as responsible for environmental documentation. Examples of the lead agency are the U.S. Environmental Protection Agency, the U.S. Nuclear Regulatory Commission, the Federal Energy Regulatory Commission, the U.S. Department of Energy, the Department of the Army, etc. Annually, more than 500 EISs are prepared and filed by the federal agencies as shown in Table 3. With rare exception, these agencies reported that the EIS process is an important aid to planning and decision making at various agency levels.[4]

When a state agency is the sole permitting/licensing authority **and there is no federal involvement,** the state agency will become the lead agency and bears responsibility for the preparation of adequate environmental documents under appropriate State Environmental Policy Acts (SEPA).

What then are the responsibilities of industries and public and private utilities in terms of environmental documentation when they initiate projects? The responsibilities are plenty and are discharged by corporate departments which are charged with the responsibility of evaluating the impacts on environment, safety, and health of corporate activities and projects. These departments prepare initial environmental evaluations, hold discussions with licensing agencies of the appropriate state and/or federal governments, prepare application for licensing which are usually accompanied by a comprehensive environmental report (ER), a safety analysis report, a technology alternatives report, and such other report required by various licensing agencies. The types of support documents prepared by industrial applicants were previously listed in Table 1.

By law, (CEQ Regulations 40 CFR 1506.5(c)) any environmental impact statement prepared to comply with NEPA shall be prepared by the lead agency or by a contractor selected by the lead agency. **The contractor shall execute a disclosure statement specifying that they have no financial or other interest in the outcome of the project.** Therefore, the role of private industry that is proposing a project and seeking licenses/permits for it would be limited to furnishing information and analysis required to prepare an environmental document by a party who can be unbiased about the outcome of the NEPA process on the implementation of the project. All information furnished by the industry is subject to technical review, quality checks, and revision as appropriate by the lead agency and its contractor.

Chapter 5

HOW TO WRITE ENVIRONMENTAL DOCUMENTS

I. INTRODUCTION

In earlier chapters, the law (NEPA) and its implementing regulations were presented. These deal with overall policies and procedures, but not with the specifics that include the scientific, technological, and environmental details of implementing NEPA. The purpose of this chapter is to introduce the scientific and technical aspects of NEPA documents. Although CEQ recognizes at least four environmental documents, the importance of NEPA seems to be derived primarily from its environmental impact statement (EIS) provisions.

A poorly prepared EIS can land the lead agency in more trouble than it bargained for. As a result, lead agencies have developed highly sophisticated operating methods and procedures for preparing these documents.

One important requirement to writing good NEPA documents is to select the right combination of professionals. Table 1 provides a list of the multidisciplinary professionals needed to prepare a fairly complex EIS. These professionals have the responsibility of sifting through conflicting data and presenting the results of their analysis of impacts in an unbiased manner to the decision maker.

The type and number of professionals needed to write environmental documents is a function of the complexity of the impacts, which, in turn, determines the level of documentation required. For example, referring to Figure 2, Chapter 4, if the proposed action is not a major federal action and has no significant effects, then an EIS will not be needed, and only an environmental evaluation (EE) will suffice. Therefore, the first step in determining the level of NEPA documentation needs is to prepare an EE. An EE helps to prepare an action description memorandum (ADM). These two documents are basic to evaluating the further need for environmental documents, but are not themselves environmental documents. If an action is listed in agency procedures as normally requiring an EIS, and in a particular circumstance the agency does not think an EIS is required, an EA should be prepared — it should not be categorically excluded.

The preliminary steps that would help to identify the level of NEPA documentation required for a proposed project can be formulated as follows:

Step 1. Study the proposed project and clearly define the action involved in the project. (When a project is proposed by an industrial applicant, getting a complete proposal from the applicant is critical to this step. This can be expedited if the applicant discusses his proposal in the formative stages face to face with the licensing agency staff before formal submission of the proposal. Such direct discussions could help in obtaining quicker feedback on the deficiencies of the proposal, thereby helping the applicant to prepare better proposals and avoid delays in implementing the NEPA process and getting the required license.)

Step 2. Complete the preliminary screening table for permits and applicable environmental regulations (see Table 2, Chapter 4)

Step 3. Identify possible permit needs and federal and state issuing agencies

Step 4. Complete the EE checklist

Step 5. Prepare an EE

Step 6. Prepare ADM (including completing the questionnaire to determine the level of NEPA documentation needed)

Table 1
TYPES OF PROFESSIONALS NEEDED IN ENVIRONMENTAL DOCUMENTATION TEAM

Air pollution control engineer
Aquatic ecologist
Marine biologist
Wildlife ecologist
Water quality specialist
Groundwater hydrologist
Socioeconomist/demographer
Process engineer
Environmental chemist
Economic impact specialist
Noise specialist
Solid waste management specialist
Meteorologist (air quality modeling)
Environmental law specialist
Project manager
Professional engineer
Health physicist
Editor
Word processor
Safety analysis specialist (accident prediction)

The reader is encouraged to actually practice these steps with the example of a proposed action provided in the NEPA compliance exercise in the following pages. Also, tables of EE checklist, permit needs, etc., as identified in Steps 3 through 6, are provided in the following pages for use by the reader.

II. NEPA COMPLIANCE EXERCISE

The six-step process can be better understood when practiced as follows, using an actual example of a large coal-fired power plant project for which the NEPA process was complied with and construction was started. In this case, the lead agency was the Federal Energy Regulatory Commission (FERC). The applicant was a public utility in South Carolina. The information below is taken from the final environmental impact statement (FEIS) on the steam plant at Santee-Cooper, South Carolina.[13]

A. Step 1. Study the Proposed Action

A public utility in South Carolina (applicant) has filed an application for approval of the use of certain lands and water for the intake and discharge system structures of the proposed steam electric generating station adjacent to the diversion canal between two lakes, as shown in Figure 1. The application was supplemented with an environmental report. The application was filed with the Federal Energy Regulatory Commission (FERC).

The proposed generating station was planned to consist of four pulverized coal-fired, steam-electric generating units with nameplate (guaranteed) generating capacities of 510 MW (450 MW net) each. The plant would be owned and operated by the applicant. The proposed plant site would be located adjacent to the diversion canal connecting lakes Marion and Moultrie. The plant would be located on a site consisting of approximately 2100 acres of presently undeveloped land owned by the applicant.

The applicant requested the FERC to issue a license for terms up to 50 years for the construction and operation of this nonfederal electricity generating project.

The major component systems of each of the four units planned at the electric power station include the following (see Figure 2):

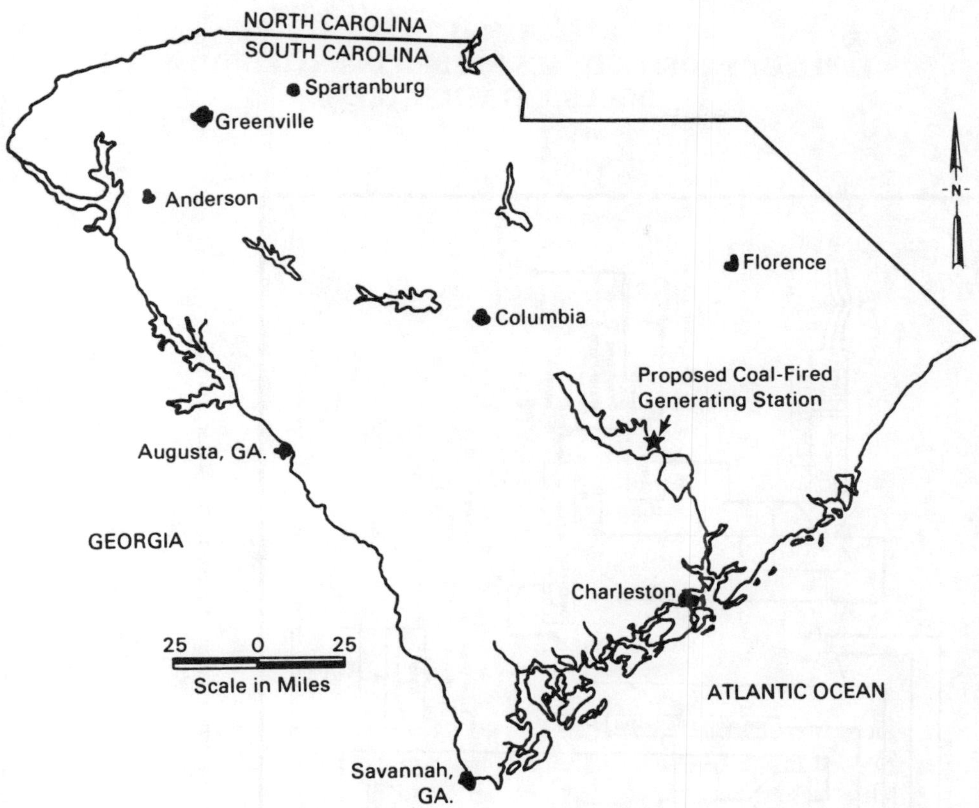

FIGURE 1. Location of coal-fired generating station.[13]

1. Steam generator
2. Turbine generator
3. Condenser and heaters
4. Mechanical draft evaporative cooling towers
5. Electrostatic precipitator in combination with a wet limestone flue gas desulfurization (FGD) system
6. Dry fly ash handling system and silos
7. Recirculating bottom ash sluice system

In addition, plant systems common to two or more units would include

1. Two 600-ft chimneys with dual flues
2. Two water intake structures and water supply pipelines
3. Water treatment system and treated water storage tank
4. Boiler feedwater treatment system
5. Bottom ash and wastewater ponds
6. Coal unloading, conveyance, and storage facilities
7. Wastewater treatment system and discharge pipe
8. Scrubber sludge dewatering, stabilization, and disposal system
9. Switchyard
10. Six electric transmission lines
11. Railroad spur and plant track loop

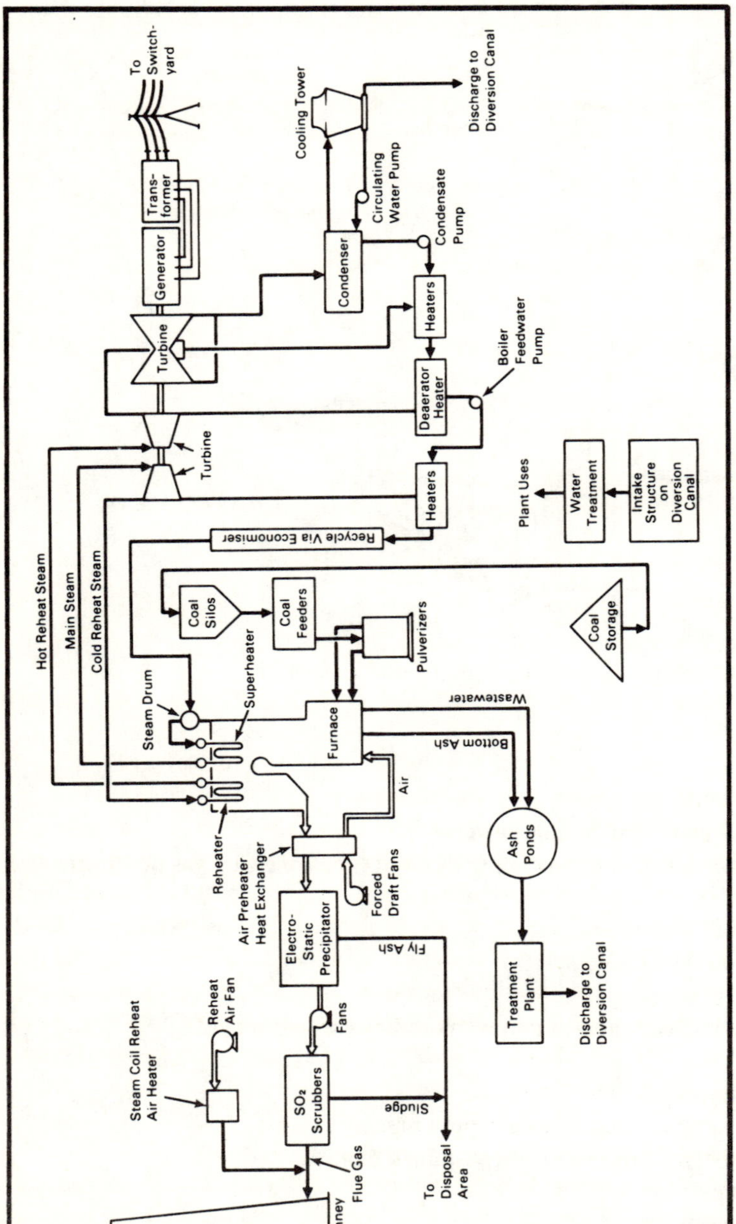

FIGURE 2. Schematic diagram of coal-fired power generating station showing environmental discharge points.[13]

Table 2
IDENTIFICATION OF FEDERAL AND STATE PERMITS NEEDED AND ISSUING AGENCIES

	Permits needed	Issuing agency
	Federal	
1.	Approval for intake and discharge system structure construction	Federal Energy Regulatory Commission
2.	Approval for intake and discharge construction	U.S. Army Corps of Engineers
3.	Notice of proposed construction of tall stacks	Federal Aviation Administration
	State	
1.	Construction permit for water pollution control facilities	DHEC[a]
2.	Construction permit for intake structure and discharge pipe	Water Resources Commission
3.	Construction permit for sewage treatment plant	DHEC
4.	Construction permit for air emission sources	DHEC
5.	Section 401 approval for intake structure and discharge pipe	DHEC
6.	National pollutant discharge elimination system (NPDES) permit for water pollution control facilities	DHEC
7.	NPDES permit for sewage treatment plant	DHEC
8.	NPDES permit for construction runoff	DHEC
9.	Prevention of significant deterioration (PSD) permit	DHEC
10.	Solid waste permit for disposal of cellulosic material by earth burial	DHEC
11.	Solid waste permit for disposal of inert, nontoxic waste by earth burial	DHEC

[a] Department of Health and Environmental Control.

The estimated construction cost of the four units is $1.6 billion in 1983 dollars, approximately.

B. Step 2. Complete the Permit Screening Questionnaire

The NEPA team should ask itself if it has sufficient data to perform the screening for permits and regulations that might apply to the proposed action. Also, is the proposed action clearly defined in terms of the duration of operation, how much fuel will be burned, what control technologies will be used, etc.? In this case, the proposed action provides sufficient preliminary data to proceed with the next step (completion of Table 2 for screening the need for permits). The reader is encouraged to try this exercise. The degree of success in completing Table 2 would be a function of the experience and understanding of coal-fired power plants and the permit process on the part of the reader. The next step is intended to help in obtaining some answers.

C. Step 3. Identify Permit Needs and Issuing Agency

In the example, the applicant would have to obtain the federal and state permits and authorizations indicated in Table 2 in order to construct, operate, and maintain the proposed project.

D. Step 4. Complete the Environmental Evaluation Checklist

Identifying needed permits is only one of the aspects of NEPA compliance, which, if rigorously evaluated, could save time in the implementation of the project. The completion of the EE checklist presented in Table 3 is an even further critical aid in understanding the

Table 3
ENVIRONMENTAL EVALUATION CHECKLIST

Quantify Potential Environmental Impacts for All Yes Answers

	Construction		Operation	
	Yes	No	Yes	No
1. **Air.** Will the proposed project/activity:				
a. Result in any gaseous discharges to the environment? (If yes, provide description, physical/chemical characterization.)				
b. Result in any particulate or droplet releases to the environment?				
c. Result in any thermal discharges to the atmosphere?				
d. Cause any other atmospheric disturbance?				
e. Violate any federal/state or local emission standards? Require PSD permit?				
f. Be subject to Federal or state standards of performance for new stationary sources? (see 40 CFR 60)				
g. Violate any applicable ambient air quality standards (e.g., CO, hydrocarbons, particulates, NO_2, etc.)?				
2. **Water.** Will the proposed project/activity:				
a. Result in any liquid discharges to the environment? (If yes, provide description, physical/chemical characterization.)				
b. Discharge heat to surface or subsurface water?				
c. Alter stream flow rates?				
d. Alter significantly natural evaporation rates?				
e. Release soluble solids to natural waters?				
f. Interconnect aquifers and impact groundwaters?				
g. Require installation of wells?				
h. Require review/permit under the Federal National Pollutant Discharge Elimination System?				
i. Require a Corps of Engineer or other permit?				
j. Violate any state water quality standards (COD, BOD, TOC, DO, TDS, pH, temperatures, etc.)?				
k. Require an Oil or Chemical Spill Control and Prevention Plan (40 CFR 110—112)?				
3. **Land Use.** Will the project:				
a. Conflict with existing zoning or land use?				
b. Be located on the 100- or 500-year floodplain (EO 11988)?				
c. Be located on wetlands (EO 11990)?				
d. Generate hazardous solid waste according to 40 CFR 206—265?				
e. Generate radioactive waste?				
f. Generate other solid waste?				
g. Impact prime or unique farmland?				
4. **General.** Will the proposed project/activity:				
a. Be subject to any other federal, state or local environmental regulations not otherwise addressed in this checklist?				
b. Increase noise levels?				
c. Impact archaeological or historic sites?				
d. Require use of carcinogens, pesticides, or toxic substances?				
e. Impact wildlife or habitat (terrestrial or aquatic)?				
f. Affect endangered species or critical habitat?				
g. Require long-term commitment of nonrenewable resources?				
h. Require new utilities or modifications to existing facilities?				
i. Increase offsite radiation dose?				
j. Impact recreation?				
5. **Legal Review.**				
a. Is there a legal history for the type of project of action proposed here?				
b. Has this type of activity been litigated under NEPA (legal cases)?				

potential environmental insult that the project could cause. The checklist covers the three media (air, water, and land) which the project can impact. It also contains questions regarding noise impacts and impacts on archeological and historic sites.

E. Step 5. Prepare an Environmental Evaluation (EE)

The outline of an EE and that of an action description memorandum (ADM) presented in the succeeding pages are very similar and apparently duplicative. Then why prepare both? The question is valid until the following distinction is clearly understood.

- The EE is prepared by the technical team at a technical level and serves as input to the ADM.
- The ADM is prepared for use at the policy level and helps determine the level of NEPA documentation needed.

In theory, the EE will permit the technical team to provide their inputs purely on a technical-scientific basis without bias. Using this input, an ADM can be prepared which takes into account not only the technical content of the EE, but also the policy nuances and legal aspects of the applicability of NEPA to the proposed project. Thus, it is the ADM that would serve as the first-level decision document.

The reader is encouraged to prepare an EE using the information developed so far on the coal-fired power plant project.

1. Outline an Environmental Evaluation

The Definition of an environmental evaluation (EE) is a document characterizing, in summary fashion, the potential environmental impact of a proposed activity. An EE is not a NEPA document, but could lead to the development of one.

The purpose of the EE is to ensure consistent documented review of environmental consequences of those activities that potentially have an environmental impact. The EE is intended to fulfill the requirements for a memorandum to files for actions not requiring an environmental assessment (EA) or an environmental impact statement (EIS) under NEPA and Council on Environmental Quality (CEQ) implementing regulations, and consists of a narrative description and an EE checklist. The EE format is as follows:

1. **Narrative** — This brief narrative section should describe the what, why, and when of the proposed action as follows:

   a. Describe the action (about $^1/_2$ page).
   b. Discuss the purpose of the action (1 page + figure).
   c. Show the location in a standard area map (1 page).
   d. State the anticipated dates for construction and operation and the planned or estimated cost of the action or the planned starting and ending dates of the project.
   e. Discuss any items marked ''Yes'' on the EE checklist.

2. **Environmental evaluation checklist** — The anticipated environmental impact of the action can be estimated, in part, using the EE checklist. Each item on the checklist should be addressed. In an accompanying narrative, the impact of those items marked ''Yes'' relative to standards, changes in environmental quality, etc. should be discussed. The checklist is intended to provide a framework for judging ''impact'' and to identify aspects of the action which may deserve additional attention. It is not necessarily complete nor is it intended to imply that these are the only items of concern.

2. Outline an Action Description Memorandum

An action description memorandum serves as the basis for a determination of the required level of NEPA documentation. It should be prepared for each proposed action as early as possible in the planning process of an action. This will help ensure proper planning for NEPA compliance and allow integration of the NEPA process with the development process for an action and thereby help avoid delays. The action description memorandum should contain sufficient information to permit a reasonable determination of the NEPA documentation required.

An action description memorandum should contain, as appropriate, the following information:

1. **Concise description of proposed action** — The proposed action should be briefly, but concisely, described. Included in this description is a statement as to the purpose and need for the proposed action; the type of action (program, policy, or project); the class of action (legislative or administrative); the type of energy technology; and an indication of the size of the action.
2. **Location of the action** — The proposed location of the action, if site specific should be identified clearly by naming the closest city or metropolitan area and the county and state in which the action is to take place. Other characteristics, such as rural or urban qualities, environmental setting (forested, desert, outer continental shelf, grasslands, etc.), and economic conditions of the area, may be included.
3. **Potential Issues** — Any known or potential issues or problems, particularly environmental issues, should be briefly presented. The results of the EE should be used here.

Notice that both ADM and EE have very similar outlines. Usually an EE can be prepared together with an ADM and the two integrated.

F. Step 6. Prepare Action Description Memorandum (ADM)

After reviewing the EE, the ADM should be prepared by an experienced environmental evaluation professional since the end result of the ADM would be to recommend the level of NEPA document required for the proposed project. As preparation to assembling the ADM, completing the questionnaire on the level of NEPA documentation (Table 4) is a requirement. When the issues related to the action are clear in the writer's mind, and the factual contents that would form the ADM are well formulated, the ADM can be prepared using the outline provided in the following pages. If the issues are not clear and factual contents are not well formulated, the ADM will not be a useful document; therefore, attempts should first be made to remedy the problems relating to issues and factual content.

In preparing the ADM, the definition of "impact" should be formulated. One such definition is provided below.

- Impact means change in the socioeconomic or biophysical characteristics of the environment which may result from a proposed action.
- The change may be beneficial or adverse from health, safety, and welfare considerations.
- The change should be quantified to determine significance of the change, and the quantified change should be related to a baseline, usually defined as the no-action or continue present action mode.
- The ADM and EE equation can be visualized as follows:

ADM = RESULTS OF EE + LEGAL, POLICY, AND PROGRAMMATIC DATA

Table 4
QUESTIONNAIRE TO DETERMINE LEVEL OF NEPA DOCUMENTATION NEEDED (ATTACH TO ADM)

Use This Sheet for Describing Your Logic for Selecting the Level of NEPA Documentation Needed

	Criteria for yes/no answer
1. Categorical exclusion?	
2. Is it a major action? (see 40 CFR 1502.3 and 1508.18)	
3. Are the expected impacts significant? (see 40 CFR 1508.27)	
4. Is there a precedent you can rely on?	
5. Is the proposed action controversial in nature?	
6. Other considerations including legal precedent, etc.?	

NEPA team recommendation____________________ Legal staff opinion____________________

III. NEPA PROJECT MANAGEMENT

The NEPA compliance exercise example described so far, when conducted effectively, leads to the identification of the level of NEPA compliance documentation required for the action described (example: the coal-fired power plant proposal). In the example exercise, the recommendation is to prepare a full-scale environmental impact statement (EIS). In instances where it is not clear that an EIS is needed, the recommendation would be to prepare an environmental assessment (EA) which determines if a finding of no significant impact (FONSI) is valid. If a FONSI is not valid, an EIS is the resulting requirement.

With the decision to write an EA or an EIS, the scope of activities immediately enlarges enough to warrant the formation of an EIS team consisting of the selected professionals identified in Table 1. This team will need a project manager.

A. Role of the Project Manager

The project manager of the NEPA team will do the following:

- Provide day-to-day technical direction
- Schedule the work and insure that schedules are met
- Control costs
- Coordinate with various departments and disciplines
- Provide overall integration of the technical, scientific, and policy aspects of the project
- Provide for quality control and peer review

The project manager's responsibilities are to motivate the team, identify the key issues, and insure that cost, schedule, and quality are maintained. Usually, for projects which are heavily engineering discipline oriented, a qualification of the project manager is that he should be a registered professional engineer. Excellent writing and communicating skills are very important. The ability to grasp the essence of the various sciences (geology, atmospheric chemistry, ecology, etc.) is indispensable if the project manager is to lead the team of such scientists.

As a matter of ethics, the project manager and the entire EIS team should remember and continually remind themselves that their analysis of the environmental-technical-economic aspects of the project is not merely a service to the decision maker who seeks help from it; it is also a service to the community. The team, in making explicit both the objectives of the action and any value judgments in the analysis, will prompt the decision maker to examine and understand the logic of his own actions. This in turn will encourage the decision maker's self-scrutiny whenever he may do something that is contrary to the welfare of the society to whom he is eventually accountable. In a sense, the NEPA process manager and team help the decision maker carry the traditional practice of requiring financial accounting of public or company funds a step further. The requirement to justify decisions that impact the health and welfare of society by explicit prior analysis should discourage hasty decisions for immediate gains, encourage some altruism and overall welfare, and lead to optimal decisions for the good of most elements of society.

B. Role of the Editor

An environmental document (EA or EIS) is much more complex than writing a technical paper or even a text book or fiction. The EIS is a public information document that is scientific and technical and very multidisciplinary in nature. There are several scientists and engineers who provide written parts of the document and who are generally not trained to write for public information, but are trained to communicate to their own kind. Also, the EIS is a public information document about actions that often affect the lives of people in significant ways, and therefore, the people deserve to receive a document that they can understand. It is here that a technical editor (technical writer), working in concert with the EIS team and, more specifically, with the EIS project manager, can act as an interpreter between the public and EIS technical staff. To fulfill this role, the editor should attend public hearings, study public comments, and try to understand public concerns as well as the project manager does.

According to Covert,[7] the public information technical writer must

1. Establish and maintain good working relationships with technical resource people.
2. Keep in mind that this is **not** public relations, therefore avoiding an advocacy role.
3. Make the maximum use of existing information.
4. Respect the often cumbersome review and revision process.
5. Listen for and use feedback from the public.

Following these five rules will not only help in doing the job properly, but will also help maintain the writer's sanity. Some EIS projects are so large requiring several editors that a lead editor will be necessary. In this instance, the lead editor will have to build the team responsible for report production by working with word processors, graphic arts staff, and duplicating/printing shop.

The lead editor should emphasize

- Quality.
- Being productive.

- Negotiating schedules and deliverables with all team members.
- Good planning.
- Avoiding bias and advocacy.
- When in doubt, do not edit technical parts without sitting with the author and getting his input.

Another important function of the editor is to help (and goad if necessary) the EIS project manager to

1. Include the editor in document planning and scheduling.
2. Include the editor in meetings likely to be informative or profitable to attend.
3. Make necessary decisions and resolve issues.
4. Communicate and inform the team.
5. Negotiate and maintain realistic schedules that allow time for editing, word processing, quality check, proofing, and correction.
6. Review material and serve as clearinghouse for contributions.
7. Help keep technical contributors on schedule and maintain quality of their contributions.
8. Obtain an EIS distribution list and other items necessary for the EIS production from the lead agency.

To sum up, the editor is a key team member in EIS projects. Powell[27] has provided an excellent answer to the question, ''How soon should editors get involved in the EIS process?'' Her answer, ''as soon as planning starts,'' best represents the author's own experience in several EIS projects. Perhaps more important than the time at which the editor joins the effort is what the editor does initially in the project. Powell gives an excellent summary of some of the things to do as follows:

> *Listen carefully* to learn needs and expectations. What principal message is this document to express clearly? In what tone? Where needs may be overlooked, *inquire*. (For instance, learn who should receive what materials for the quality assurance file. Such items need to be carefully spelled out.) Inform the program manager of what editors can do to help the project proceed more smoothly. There are many things editors can achieve early, but focus your efforts.
>
> With outlines and early drafts, it's important to review for consistent parallel order of elements (e.g., action alternatives in an EIS) in the various sections and in tables, etc. Determine this order and establish it early. Are there enough headings to guide the reader clearly? To what level will headings be numbered?
>
> - Establish ground rules together with word processing:
>   - — Key words
>   - — Any special terms
>   - — Acronyms
>   - — General format (order within sections)
>   - — Table format (headings, order of topics, etc.)
>   - — Punctuation
>   - — Units of measurement sponsor requires (any exceptions?)
>   - — Indexing
>   - — References
>   - — Other special requirements
> - Determine early what the eventual format will be and inform word processing:
>   - — Consult with word processing in choosing the format. Be sure it is sufficiently flexible as to page lengths, etc., to accommodate eleventh-hour changes.
>   - — Stay in draft format long enough. It permits more accurate proofing. We avoid single-spacing where possible: it is most difficult to edit, revise, proof, and read. Also, two-column may be impractical and expensive for documents that will be reviewed numerous times. Two-column is good — and cost-effective — for publications less subject to change.

Table 5
CONTENTS OF AN EIS IMPLEMENTATION PLAN

Abbreviations and Acronyms
1.0 Introduction
2.0 Purpose and Background
- 2.1 Purpose of Environmental Impact Statement
- 2.2 Background

3.0 Scope
- 3.1 Proposed Action
- 3.2 Alternatives to the Proposed Action
- 3.3 Excluded from Detailed Consideration
- 3.4 Topics Included in EIS from Scoping Process
- 3.5 Identification of Significant Environmental Issues
- 3.6 Documents Related to, but Not Part of, the Scope of this Environmental Impact Statement
- 3.7 Environmental Reviews and Consultation Requirements

4.0 EIS Preparation
- 4.1 Tentative Decision Making Schedule
- 4.2 Proposed Schedule for EIS Preparation
- 4.3 Project Management
- 4.4 Draft EIS Outline

5.0 EIS Scoping Process
Appendix A — Notice of Intent as it Appeared in *Federal Register*
Appendix B — Comment Letters Received on the Notice of Intent

IV. EIS IMPLEMENTATION PLANT (IP)

The implementation plan (IP) is another document that plays a very important role in the NEPA process, but is not defined as a requirement by the CEQ regulations. The IP should be written only if an EIS is to be prepared. The suggested contents for an IP are described in Table 5.

The IP does not contain cost information, unlike a work plan. Two key elements that the IP should contain are (1) tentative decision making schedule for the project and how it relates to the NEPA process, and (2) a detailed outline of the draft EIS.

V. TECHNICAL APPROACH TO EA/EIS PREPARATION

This book is about the NEPA process methodology, i.e., how to comply with the act. It is not a technical treatise on how to model, calculate, and quantify environmental impacts from various industrial sources. There are over 100 volumes available that describe the various methods to calculate the environmental impacts from emissions from the numerous industries that man has invented and is using.

Yet, it is important to get a general idea of the technical aspects of the NEPA process. The step-by-step technical approach is illustrated in this portion of this chapter, ''How to Write Environmental Documents''.

The preceding discussions outlined the methods for writing the EE and the ADM. They are simpler than writing EAs and EISs. The following discussion would apply to both the EA and the EIS.

To write the EIS, the first step is to describe the proposed action as clearly and quantifiably as possible. This is the first of the five major steps involved in the technical approach EA/EIS writing, as shown in the approach flow diagram in Figure 3.

A. Describe the Proposed Action

Describing the proposed action in the early stages of a project is not easy because the conceptual design team is still conceptualizing the project. Usually, the EIS team acts as a

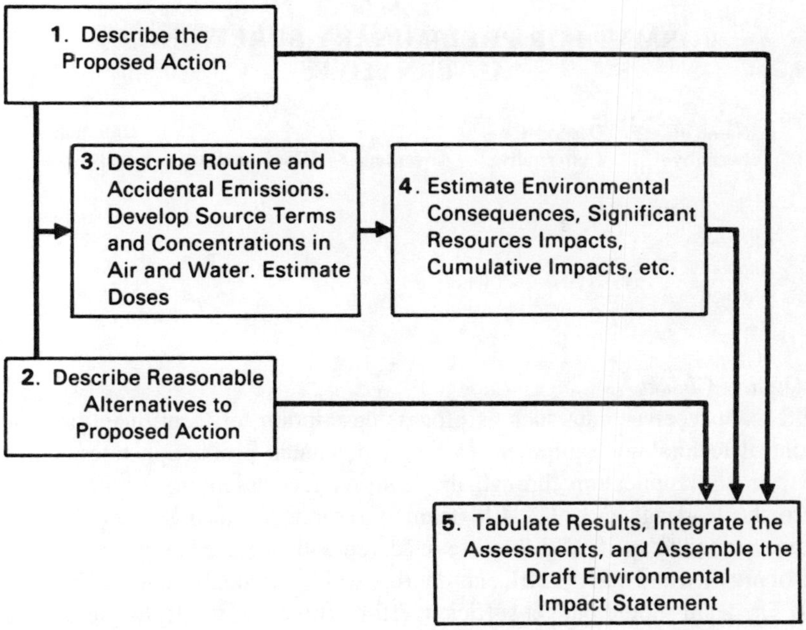

FIGURE 3. Technical approach flow diagram for EA and EIS preparation

catalyst to the conceptual design team, goading them to clearly define the action so that the following items are clearly quantified.

1. What is the objective and annual production rate of the project?
2. What are the raw material inputs?
3. How long will be plant life?
4. How many hours will it operate per year?
5. What are the emission points?
6. What amounts of gaseous, solid, and liquid effluents will be expected?
7. What are the compositions of the effluents?
8. How will the effluents be disposed of?
9. What control technologies will be employed to reduce effluents?
10. Will process modifications be considered as a method of controlling or eliminating effluents?
11. What is the estimated cost of construction?
12. What emissions are expected during construction?
13. Where will the plant be located?
14. How many people will be employed during construction and during operation?
15. What is the estimated cost of operation annually?
16. How much resources (water, electricity, petroleum, etc.) will be used annually?

As a minimum, the proposed action should describe the following major items in detail:

- Facility description
- Process description that the facility will use
- Construction and operation schedule
- Pollution control systems
- Raw material usage
- Effluent volumes and composition

Table 6
FORMAT FOR PRELIMINARY SCREENING OF ALTERNATIVES

Name of alternative	Description of alternative	Advantage	Disadvantages	Rationale for dismissal
Alternative 1				
Alternative 2				
Alternative 3				
Alternative 4				
(etc.)				

1. Data Quality Checks

The EIS team receives data such as process description, effluent quantities and composition, control technology equipment to be used, annual production rates, and resources required from the application through the lead agency (assuming the EIS team is under contract to the lead agency). The EIS team is responsible to make certain that the data received are audited to verify that they are in accord with accepted engineering and scientific standards of practice and that the effluents data are within reasonable range of best engineering estimates. The team should budget sufficient efforts (time and funds) to check the basic data for accuracy, reliability, and defensibility in a court of law if required. If this data quality audit step is not carefully carried out, the entire EIS may become flawed and may have to be redone at great cost, time, and loss of reputation. Therefore, the best available experts should be employed to audit the applicant's data; obviously, experts employed should have nothing to gain from the outcome of the EIS process.

B. Describe Reasonable Alternatives

It is not often realized that the selection of alternatives to achieve the purpose and need that would be satisfied by the proposed action is very dependent on the need for the proposed action. Therefore, before selecting alternatives, there should be a clear statement of need prepared and circulated among the EIS team. For example, in the exercise described in Section II of this chapter, the need to which the proposed action addresses is the need for electric power. One of the ways in which this need can be met is to construct the four units of 510-MW coal-fired power plants. The alternatives to this proposed power plant could be listed as follows, since these appear to meet the need for power:

1. Upgrade the older power plants (if any).
2. Purchase power from other sources.
3. Conserve power by load management and instituting a conservation program.
4. Construct a nuclear generating plant to reduce pollution from emissions during coal burning.
5. Construct to allow oil/gas as alternative fuels.
6. Consider fluidized-bed coal combustion instead of conventional boilers.

These are some of the apparently reasonable alternatives which, upon closer examination, may not be found to be reasonable. The purpose of this exercise in looking at alternatives is to make certain that the proposed action is not selected without due and deliberate consideration of better methods that may be available to achieve the goals of the proposed action.

Before a set of alternatives are selected for detailed examination, it helps to prepare a table of alternatives, as shown in Table 6.

For those alternatives not dismissed by this procedure, a detailed description will have to be prepared using the same type of data as for the proposed action. Adequate care should be taken to insure that the EIS team is serious about studying the alternatives to really identify the optimum method of achieving the purpose and need, rather than to perfunctorily satisfy the CEQ requirements. Equal weight should be given to evaluate the alternatives as to the proposed action so far as possible.

1. Guidelines for Alternatives Analysis

To say how preferred an alternative is in comparison to others and why, the analyst of the alternative(s) needs a criterion for ranking them and a way to measure the significance of each of their consequences. The analyst's role here is not to say what the decision ought to be; his aim is to rank the alternatives based on the particular criterion chosen. It is appropriate in an EIS to recommend a preferred alternative that the decision maker may think about, although when an analyst recommends so, he is no longer purely an analyst. He assumes the role of an advocate; therefore, he should clearly identify where his own values have been introduced into the analysis, thereby helping to avoid value judgments and to carefully reduce bias.

Selecting criteria for ranking alternatives is not an easy task. Even if the criteria are wisely selected, applying them to the ranking process is equally difficult. Examples of criteria for ranking alternatives are

1. Potential environmental impacts
2. Status of technology
3. Timeliness of meeting the need
4. Proximity to population centers
5. Extent of natural resources used
6. Dollar costs of implementing the alternative
7. Cost vs. benefit and "reasonable" cost/benefit ratio
8. Potential socioeconomic impacts
9. Impact of delays in meeting the need to be satisfied by the proposal
10. Trade-offs among the above

The selection of an alternative by the decision maker is a technical-sociopolitical-environmental exercise. The cost-benefit-risk method of decision making must be supplemented often by the personal qualities of the decision maker. Since most decisions are expensive, time consuming, irreversible, and very demanding of the qualities of the head and heart of the decision maker, the quality of the decision is a direct function of the decision maker's quality. The decision analysis method is only an aid, a tool in the hands of the decision maker, and is not a substitute for the decision maker.

Quade[29] presents a beautiful example of the difficulties of decision making which is summarized below:

The contribution of automobiles to the problem of controlling photochemical smog in southern California was analyzed by Quade's colleague Haagen-Smith who showed that unburned hydrocarbons (HC) and oxides of nitrogen (NO_x) were both required for the photochemical reactions which produced the smog. It was decided that controlling hydrocarbons would be easier. At the same time, some other workers decided that carbon monoxide (CO) emission levels needed reduction to reduce the health problems associated with CO. But CO had no connection with smog. Regulations were promulgated to curb both CO and

HC from automobiles. The natural response of the automobile industry was to raise the flame temperature in the engines to burn the hydrocarbons more completely, which unwittingly increased the emission of NO_x, thereby cancelling the benefit of the regulation of hydrocarbons. When the increase in car population was also included, the smog problem was not reduced at all.

This example typifies all the pitfalls of decision making: the expense of changes in engine temperature, the expense of enforcing regulations by testing car emissions, the time-consuming aspects, the irreversibility of the decision, at least until its ineffectiveness was clear and the whole regulatory process was repeated, and the study needed to devise new solutions.

In this book we can only present the difficulties of the decision making process. Devising solutions to these difficulties in your particular project is left to you, the reader. The analytical methods for decision making that Quade[29] outlines in his book, *Analysis for Public Decisions,* may be helpful to you in your efforts. The EIS writer is primarily an analyst. He works for a client on contract. The client is the decision maker. The analyst's role is not easy. The following thoughts by Sugden and Williams,[32] quoted by Quade,[29] are important for the analyst to meditate upon:

1. The analyst must own some allegiance to the cause of intellectual honesty. Thus, for example, he should not accept uncritically a decision maker's stated objective if this objective is inconsistent with policies to which the decision maker is committed. As a further aside, it should be said that the analyst, as a citizen, may properly have views about the goodness or badness of the objectives postulated by his clients. As a free man, he has the right to refuse offers of employment that would require him to use his skills in ways that he believes to be wrong. But to accept the role of analyst is to agree to work with the client's objectives.[32]
2. The role of the analyst is to assist, not simply a decision maker, but a decision making process that has the assent of the community as a whole. In this process, the community, as well as the decision maker and the analyst, is involved. The decision maker is responsible for making a decision, according to his own lights, but he is responsible to the community. His right to decide stems from the consent of the community, expressed through the political system. The community, then, ought to have the right to call upon the decision maker to account for his decisions.[32]

Despite these somewhat contradictory views about the analyst's role, he must recognize that his responsibilities include and extend beyond the interest of the sponsor and act accordingly, but after full discussions with the sponsor. The EIS worker is also a policy analyst or a policy scientist. In this role, he must press for the good and for what he believes in as the result of his analysis. But there is no hard-and-fast rule to tell him how much. One reason why policy analysis cannot be completely objective is that the analyst may have to compromise, to allow some bad to get the good accepted. Thus, he is not completely objective, but an advocate.

I would like to conclude this complex subject with a summary of the code of ethics for policy scientists. Since policy analysts have professional responsibilities, a code of ethics is desirable. Such a code may help them handle difficult values, dilemmas, and ethical issues they are likely to have to face. The Operations Research Society of America suggests a code of ethics for policy scientists in the field of applied science and engineering (Quade).[29] This code is summarized below.

SUMMARY OF CODE OF ETHICS FOR POLICY SCIENTISTS SUGGESTED BY OPERATIONS SOCIETY OF AMERICA

1. A policy scientist should not work for a client whose goals and values, in the opinion of the policy scientist, contradict basic values of democracy and human rights.
2. When the goals and values of a particular client contradict basic beliefs of the policy scientist, the policy scientist should resign, rather than help in the realization of goals and values with which he intensely and fundamentally disagrees.
3. The purpose of policy scientists is to help in better policy making, and not to displace legitimate policy makers and decision makers with policy scientists who become ''gray eminences''. Therefore, policy scientists shall try to preserve and increase the choice opportunities for their clientele, e.g., by always presenting all reasonable alternatives. In particular, a policy scientist should not hide an alternative because it contradicts his own personal values and preferences.
4. Policy scientists should explicate assumptions and should present clear value sensitivity analyses, so as further to increase the judgment opportunities for thier clientele.
5. A policy scientist should refuse to prepare studies, the sole purpose of which is to provide a supporting brief to an alternative already finally decided upon for other reasons and considerations by his client.
6. Policy scientists should not work for clients who do not provide necessary access to information and opportunities for presentation of studies and their findings.
7. All forms of conflict of interest should be avoided, including utilization of information for private purposes and presentation of recommendations in respect to subject matters in which a policy scientist has a personal and private interest.

One final piece of advice to the analyst: try to keep your work from being misinterpreted or being used to imply more than it should. Otherwise, the analysis itself may be brought into disrepute. With this advice, let us launch into estimating environmental impacts which also requires the application of all the guidelines described above.

C. Estimate (Calculate) Environmental Impacts

This part of the EIS work is the heart of the NEPA process and involves the following steps for both the proposed action and alternatives.

1. Describe routine emissions during construction and operation.
2. Describe potential operational accidents.
3. Develop source terms for routine operations and accidents.
4. Estimate ambient air concentrations that would result from the emissions.
5. Estimate water quality impacts from liquid effluents.
6. If the proposal is a nuclear facility, estimate radiation doses to the public that would result from radioactive emissions (if any).
7. Describe cumulative impacts.

Estimating environmental impacts and describing the environmental consequences requires substantial background data for the proposed site of the facility (Table 7). For most areas of the U.S., such data are available in existing environmental impact statements filed with the EPA and also in the state environmental protection departments. Another important source of background data is the ER (environmental report) that industrial applicants prepare and submit to the licensing agency with the license application.

A frequently asked question is what to do when the background data or design data available are inadequate for performing a complete analysis. This problem is not unusual

Table 7
EXAMPLES OF BACKGROUND DATA NEEDED FOR CALCULATING AND COMPARING ENVIRONMENTAL CONSEQUENCES

- Air quality levels
- Water quantity levels
- Population data levels in all media
- Ecological resources
- Endangered species
- Aquatic fauna
- Noise levels
- Community services (police, fire, schools, etc.)
- Archeological and historical points of interest
- Meteorological data
- Diffusion factors
- Models to calculate pollutant
- Radiation dose computation models
- Population distribution models

because the timing of NEPA documents is such that operational data or design estimates are not fully available. In such circumstances it should be remembered that the purpose of NEPA is to foster excellent decisions, not to generate definitive scientific documents. When federal agencies lack definitive information to analyze the environmental effects of high risk activities well into the future, they are required to (1) fully disclose the lack of important information specifically and (2) engage in reasonable forecasting about the likely consequences of proceeding without such information. Unequivocal proof of compliance is neither expected nor required because of the substantial uncertainties inherent in long-term projections. This may be called the "rule of reason" and it would encompass a "bounding analysis" based on available data and, in some cases, engineering estimates. Recently, the CEQ also has rescinded the "worst case" approach (40 CFR 1502.22) and returned to the rule of reason approach. A bounding analysis (usually performed in the preparation of a complex EA and EIS) often occurs early in the NEPA process. There is no universally accepted definition of bounding analysis. The following definition should be helpful as a guide.

A bounding analysis is one that is performed with the available data, modeling assumptions, and accidental release scenarios, the total of which is sufficiently conservative so that there is high confidence that, as a result of the *compounded conservatisms,* the calculated (predicted) environmental impacts will exceed those to be expected in actual practice. Individual parameters selected may not necessarily be extreme values and may be mean values in cases where sufficient data are available to warrant such a choice of values well on the conservative side of the expected mean. Modeling assumptions, in the face of uncertainty, may be made on the side of conservatism. Finally, accident scenarios may be chosen based on their credibility, to describe the most serious incidents that could reasonably occur. (The reader is welcome to refine this definition of bounding analysis.)

In contrast to a bounding analysis, if a worst case analysis were performed, the parameters chosen would be extreme values. Most often bounding analysis is applicable to estimating impacts of unplanned events (accidents) at nuclear facilities and during nuclear transportation. The nuclear facility accidents could be postulated to occur either within the plant or due to events from outside the plant (airplane hit, earthquake, sabotage, etc.).

When NEPA documents include description of accidents, the probability values for the occurrence of the accidents should also be included, as a rule. This is particularly important when the subject of the NEPA document is a nuclear power plant, a defense nuclear material production facility, or a nuclear waste storage and disposal project. Probabilities are useful to estimate risk defined as

Table 8
QUALITATIVE UNDERSTANDING OF PROBABILITY NUMBERS

Probability	Qualitative interpretation
1.0	Very likely or certain, or unknown (for conservatism)
10^{-1}	Less than certain, but reasonably likely
10^{-2}	Not likely, but cannot be ruled out
10^{-3}	Probably will not occur (frequently used to indicate that the available data show that the event or process will not occur, but that the available data are questionable)
10^{-4}	Very unlikely, based upon reliable data
10^{-5}	Extremely unlikely
10^{-6}	Physically possible, but almost certain not to occur
0	Assumed to be physically impossible, based on the currently available data

Table 9
DESCRIPTION OF PROBABILITY SCALE[16]

Probability scale	Description	Estimated range of probability of accident occurrence, per year
E (Extremely high)	Likely to occur one or more times per year	$P \sim 1.0$
D (High)	Likely to occur once every 10 years	$10^{-1} \leqq P < 1.0$
C (Medium)	Likely to occur once every 100 years	$10^{-2} \leqq P < 10^{-1}$
B (Low)	Likely to occur once every 1000 years	$10^{-3} \leqq P < 10^{-2}$
A (Extremely low)	Likely to occur less than once every 1000 years	$P < 10^{-3}$

- risk = probability × consequences

In nuclear facilities, for example, the consequence of an accident is the radiation dose (expressed as rem) to be received by the general public. The dose may be multiplied by the probability to provide a dose risk, not a widely used concept. A more useful concept is to convert the dose to estimated health effects (fatal cancers or specific genetic effects). Even if the dose be high, but the probability be low (<0.001 or 10^{-3}) or extremely unlikely ($<10^{-5}$), then the risk would perhaps turn to be also low or extremely unlikely. Man lives in a world of risks which he accepts as a necessity when he feels that the benefits outweigh the risks. The important point is that the projected risks of a proposed project should compare favorably with the risk-benefit pattern of other activities, such as driving an automobile or flying in an airplane. The estimation and use of risk and probabilities is not an exact science, but they are the best available measure of health hazards to the public. Tables 8, 9, and 10 may provide some guidance in interpreting probabilities and risks.

1. Individual Impacts Estimation

Table 11 provides further details of items to be covered under the various key environmental media, such as

Table 10
RISK FACTORS GENERALLY USED TO INDICATE HEALTH EFFECTS RELATED TO NUCLEAR FACILITIES OPERATIONS[a]

Type of risk	Predicted incidence per 10^6 person-rem[b]
Fatal cancers from:	
Total-body exposure	50 to 500
Lung exposure	10 to 100
Bone exposure	1 to 5
Thyroid exposure	1 to 5
Specific genetic effects to all generations from total-body exposure	50 to 500
Total	100 to 1000

[a] This table presents general information only. The interpretation of health effects and risks is the subject of several documents (NCRP 1975, U.S. DOE 1980) which should be referred to for fuller understanding.

[b] Person-rem = total population (persons) × radiation dose (rem). One million person-rem is obtained by multiplying the population (say, 500,000) by the estimated dose (say, 2 rem). This one million person-rem could result in 50 to 500 fatal cancers in the population plus 50 to 500 genetic effects, or a total of 100 to 1000 health effects per 10^6 person-rem total population dose.

- Air medium
- Water medium
- Land medium
- Ecology
- Sound and noise
- Human aspects
- Economics
- Resources

Several detailed descriptions of the approach to evaluating environmental impacts are available in published literature.[14,17,25]

Ortolano[25] discusses methods for estimating air quality impacts in a straightforward presentation of six steps. This method is a good example of how to establish the air quality impacts of a proposed project. The six steps are

1. Establish background air quality levels for air quality indicators (pollutants) likely to be influenced by the proposed project.
2. Identify applicable air quality criteria and standards including PSD requirements.
3. Forecast future air pollutant emissions (in weight per unit time or grams per second) with and without the project.
4. Forecast future ambient air pollutant concentrations (using appropriate air quality diffusion models and climatological inputs) with and without the project.
5. Compare predicted air quality (from operation of the project) with applicable standards and PSE regulations and increments.
6. Modify plans as necessary to deal with potential air quality problems (include improvements that mitigation procedures offer).

Table 11
LIST OF IMPORTANT ENVIRONMENTAL FACTORS TO BE ADDRESSED IN ENVIRONMENTAL CONSEQUENCES

Air quality
1. Radioactivity
2. Particulates
3. Sulfur oxides
4. Hydrocarbons
5. Nitrogen oxide
6. Carbon monoxide
7. Ozone
8. Hazardous toxicants (lead, etc.)
9. Odors

Water quality

10. Aquifer safe yield
11. Flow variations
12. Oil
13. Radioactivity
14. Suspended solids
15. Thermal pollution
16. Acid and alkali
17. Biochemical oxygen demand (BOD)
18. Dissolved oxygen (DO)
19. Dissolved solids (TDS)
20. Nutrients
21. Toxic compounds
22. Aquatic life
23. Fecal coliforms

Land

24. Soil stability
25. Archeological/historical sites
26. Land use patterns
27. Existing rights of the Indian nation

Ecology

28. Large animals (wild and domestic)
29. Predatory birds
30. Small game
31. Fish, shell fish, and water fowl
32. Field crops
33. Threatened species
34. Natural land vegetation
35. Aquatic plants

Sound

36. Physiological effects
37. Psychological effects
38. Communication effects
39. Performance effects
40. Social behavior effects

Human aspects

41. Life styles
42. Psychological needs
43. Physiological systems
44. Community needs

Economics

45. Regional economic stability
46. Public sector review
47. Per capita consumption

Resources

48. Fuel resources
49. Nonfuel resources
50. Aesthetics

Ortolano[25] also provides a very good introduction to models for calculating ambient air quality concentrations. Some simple models he describes are box models, rollback models, statistical models, and combination rollback and statistical models. These simple models are usually not adequate to satisfy the requirements of environmental impact statements, but can be used as rough indicators of how emissions may affect air quality. Rigorous estimates of air quality impacts can be conducted using atmospheric dispersion models such as (1) point source Gaussian diffusion models, (2) multiple-source models for chemically stable substances, (3) models for unstable (radioactive) substances from point and multiple sources, and (4) models for chemically active substances.

No matter which model is used to calculate air quality concentrations, care should be taken to ensure that the model used is one approved by the U.S. EPA. It is better to discuss the model to be used with the EPA before calculations begin. Also, most often, forecasting pollutant concentrations either in air or water is best left to the air quality or water quality modeling specialists whose job it is to ensure that approved models are used.

In discussing the environmental consequences of nonnuclear facilities, the following items should be evaluated and results presented in a manner essential to the decision making process:

1. Exposure of the general public and operating personnel to pollutants emitted during normal operation
2. Changes in air and water quality as a result of normal operation
3. Construction-related impacts
4. Changes in short- and long-term land use
5. Socioeconomic impacts on surrounding communities
6. Irretrievable and irreversible commitments of resources
7. Measures to mitigate impacts
8. Compliance with federal, state, and local regulations as applicable, including compliance with PSD requirements
9. Cumulative impacts

If the environmental consequences of nuclear facilities with radioactive emissions are to be discussed, the following four items should be added to the nine items identified above.

1. Exposure of the general public and operating personnel to radiation from emissions during normal operation
2. Exposure of the general public to radiation from emissions during abnormal operation (accidents)
3. Radiation exposure to the general public and operating personnel from transportation of nuclear material, if any
4. Decontamination and decommissioning

For the coal-fired power plant case described before, the environmental consequences analysis will be presented for SO_2 and particulates as an example for the reader's benefit. This example shows how to measure impacts, describe the impacts, and compare them to existing standards.

The first step in estimating the environmental consequences of SO_2 and total suspended particulates (TSP) is to recognize that the source of these two pollutants is the combustion of coal. The sulfur content of coal, the ash content of coal, and the quantity of coal burned together determine the source term (subject to the control technology to be employed).

The following analysis of air quality impacts are extracted from the Final Environmental Impact Statement FERC/EIS-0013, July 1980, published by the Office of Electric Power Regulation, Federal Energy Regulatory Commission.

The analysis is presented on the following assumptions and estimates:

1. Coal Characteristics

	Average	Maximum
Sulfur content (%)	1.8	2—3
Ash content (%)	16.5	20
Heating value (BTU/lb)	11,200	10,800

2. Heat rate (electricity generation efficiency) is 9750/KWh.
3. The boiler is assumed to be operated at the maximum operating level of 530 MW per unit or 104% of rated capacity. This obviously a theoretical case to bound the impacts.
4. The coal burning rate corresponding to 530 MW power generation/unit/year would be 230 ton/hr of average coal composition and 238.5 ton/hr of maximum coal composition.
5. The assumed efficiency of the SO_2 control equipment is 90%; that of the particulate control equipment is 99.7%.
6. The estimated SO_2 emission rate would be 1656 lb/hr (average coal) and 2194 lb/hr (maximum sulfur content coal).

Table 12
PARAMETERS USED IN ATMOSPHERIC MODELING FOR EACH OF FOUR POWER PLANT UNITS

Expected maximum load (MW) per unit	530
Fuel consumed (10^6 Btu/hr)/(g/sec)	5,200/59,000
Stack height (ft)/(m)	600/183
Stack diameter (ft)/(m)	31/9.5
Stack exit temperature (°F/K)	140/333
Stack exit velocity (ft/sec)/(m/sec)	72/22
SO_2 control efficiency, units 1 & 2 (%)	85
SO_2 control efficiency, units 3 & 4 (%)	90
TSP control efficiency (%)	99.8
SO_2 emission rate, units 1 & 2 (lb/hr)/(g/sec)	3,100/390
SO_2 emission rate, units 3 & 4 (lb/hr)/(g/sec)	2,100/260
TSP emission rate (lb/hr)/(g/sec)	160/20
NO_x emission rate (lb/hr)/(g/sec)	3,200/390
Location (UTM coord., km)	582.8 E, 3691.6 N

Table 13
POLLUTANT CONCENTRATIONS PREDICTED BY AIR QUALITY MODEL (μg/m³)

Pollutant	Averaging period	Maximum impact[a]	Existing level[b]	Total[c]	State standard	Federal standard
SO_2	3 hr	289	68	357	1300	1300
	14 hr	38	22	60	365	260
	Annual	1.8	5.4	7.2	80	80
TSP	24 hr	2.4	118	120	250	250
	Annual (GM)	0.1	51	51.1	60	60

[a] Predicted ambient concentration by model from all four units.
[b] Existing background concentration in ambient air.
[c] Total ambient concentration, sum of a and b.

7. The estimated TSP emission rate would be 228 lb/hr (average coal) and 286 lb/hr (maximum coal).

Items 6 and 7 provide the source terms for air quality modeling. Four air dispersion models, developed and approved by the EPA, were used to predict the impact of the power plant units and local existing industrial sources on local air quality. The PTMAX model was used to establish downwind distances of maximum impact. The single-source (CRSTER model was then used to predict both short-term (24- and 3-hr peak impacts) and annual-average impacts. The CRSTER modeling utilized 6 years (1964 and 1970 to 1974, inclusive) of hour-by-hour meteorological data. The interactions between the proposed plant and other major point sources in the vicinity were evaluated with the PTMTPW model, using data obtained from the South Carolina Department of Health and Environmental Control. The CDM model evaluated annual-average cumulative impacts at the proposed plant, at Cape Romain, and at the North Charleston nonattainment area (NAA), i.e., an area in which the ambient air quality standards have not been attained.

The modeling calculations were based on the assumed parameters as input data shown in Table 12. The data in Table 12 show approximately the types of information needed in performing environmental consequences analysis relative to the ambient air.

In Table 13 are presented the maximum expected ambient concentrations of SO_2 and TSP

Table 14
PSD INCREMENT COMPLIANCE DATA

		Class II area ($\mu g/m^3$)		Class II area ($\mu g/m^3$)	
Pollutant	**Averaging period**	**Maximum allowable increment**	**Four units maximum impact**	**Maximum allowable increment**	**Four units maximum impact**[a]
SO_2	3 hr	512	289	25	18.3
	24 hr	91	38	5	4.9
	Annual	20	1.8	2	0.3
TSP	24 hr	37	2.4	10	0.3
	Annual	19	0.1	5	0.1

[a] At Cape Romain.

predicted by the models and compared with the state and national ambient air quality standards. The impact of the proposed facility would comply with the standards according to this table. However, will the PSD requirements be met? This question must be answered to more fully address the air quality impacts of SO_2 and TSP.

The analysis of compliance with PSD increments showed that the other major sources in the vicinity of the proposed facility, which have been constructed or modified recently, have less interactive effect with the proposed units than they have alone. Accordingly, the impacts associated with the proposed units can be compared directly to the PSD increments. Class II increments apply to the vicinity of this site; Class I increments apply to Cape Romain (about 42 mi southeast). Estimated impacts and allowable increments (compared in Table 14) indicate that compliance with PSD requirements would be achieved with two units at 85% SO_2 removal and two at 90% removal. The maximum impacts would occur closer to the site than to the Cape Romain area. The long-range impacts presented were calculated with the CRSTER model. However, these values are considered conservative because a straight-line plume path was assumed as opposed to a more realistic sector average approach. Such an approach would result in lesser predicted impacts than those presented in Table 14. Once again, the parameters are chosen to bound environmental impacts.

Another air quality impact consideration to be addressed is the compliance with nonattainment area (NAA) requirements if there are nonattainment areas impacted by the proposed facility. For SO_2, there are not nonattainment areas in South Carolina. For TSP however, there is an NAA close to site (35 mi south) near Charleston. The maximum impact of the proposed units on this NAA is contrasted below with "significant" levels defined by EPA:

		Concentration ($\mu g/m^3$)	
Pollutant	**Time period**	**Significant level for NAA**	**Four units maximum impact**
TSP	24 hr	5	0.4
	Annual	1	0.1

Since the data show that the impact of the units is not "significant", no emissions effect considerations need to be made prior to approval of the licensing application. The long-range impacts presented were calculated with the CRSTER model, considered to provide conservative values.

The preceding example of estimating environmental consequences of pollutants is meant to serve as a guide to the quality and depth of analysis needed in determining the impacts of a proposed action. Similar analysis will be required for all pollutants to the air, water, and land media.

2. Cumulative Impacts Estimation

An area most often ignored by EIS writers is the cumulative impact discussion. If it is not ignored, it is usually treated perfunctorily with just a few qualitative paragraphs to show compliance with the CEQ requirements. Cumulative impacts are to be treated as respectfully as direct and indirect impacts. No doubt, it is easier to quantify direct impacts; quantifying indirect and cumulative impacts requires a lot more additional analysis and is therefore usually treated qualitatively. This is not acceptable, particularly if the potential for the cumulative impacts to harm public health or welfare is real.

According to the CEQ regulations (40 CFR 1508.7), ''cumulative impact'' is the impact on the environment which results from the incremental impact of the action when added to other past, present, and reasonably foreseeable future actions, regardless of what agency (federal or nonfederal) or person undertakes such other actions. Cumulative impacts can result from individually minor, but collectively significant, actions taking place over a period of time.

Another way to define cumulative impacts is two or more individual effects which, when considered together, are considerable or which compound or increase other environmental impacts.

The effect of an individual project may be negligible, or can be made negligible through mitigation including use of control technologies. But the combined effect of this project with other projects could exceed the threshold at which significant environmental damage could become evident.

Dickert and Tuttle[9] describe two approaches to review projects so that cumulative impacts remain below threshold levels: (1) the incremental approach and (2) the comprehensive approach. Their definition of these two approaches is summarized below:

> In the incremental approach, projects are reviewed on a case-by-case basis on their merits without reference to other projects and without a prior knowledge of the system capacity. Case-by-case review continues until some impact becomes evident. At this point, the options are to halt further projects, ignore the impact and continue project approvals, or in some way modify additional projects to avoid the impact. This obviously is not an acceptable method.
>
> In the comprehensive approach, extensive field testing establishes the capacity of the system before project approvals are given. The rate or total amount of development is managed to stay below prestated threshold levels, and halted when such thresholds are reached. While this represents a logical approach, it is fraught with the technical difficulties of establishing system thresholds and total system capacity.

Each EIS team will have to develop a clearly defensible methodology and approach for calculating the cumulative impacts of projects in the region where the project is to be sited. Every effort should be made to ensure that the capacity of the ecosystem and allowable levels of pollutants, including radiation-dose-inducing radionuclides, in the air and water will not be exceeded now or in the future. This assurance is only possible when serious attention is paid to estimating cumulative impacts as accurately as possible. (Note: development of a comprehensive methodology for estimating cumulative impacts for use in the NEPA process could be an excellent subject for a Ph.D. dissertation.)

VI. ASSEMBLING THE EA/EIS

Using the technical approach described in Section V of this chapter, all the elements of the EA/EIS can be prepared by the various professional groups of the EIS team. The EIS project manager can now assemble the EIS/EA according to the format presented in Table 15. The EA/EIS editor and project manager need to work together to insure that the assembled document flows well and provides the information needed by the decision maker. The wise project manager will remember the following 13 rules in assembling the EIS:

1. See the trees and the forest.
2. Be detached from the outcome.
3. Be unbiased.
4. Be aware of all the relevant factors.
5. Be professional.
6. Be scientific.
7. Be to the point.
8. Do not be too brief.
9. Do not be too wordy.
10. Make every word count.
11. Always think of a high school graduate who is reading your EIS.
12. Be responsive to NEPA spirit.
13. Be perfect, it never hurts!

The document is now ready for internal review by peers and senior management.

VII. THE REVIEW OF NEPA DOCUMENTS

Both the LEIS and normal EIS have two review cycles:

1. Internal review cycle to insure senior management involvement
2. External review cycle to insure public participation

A. Internal Review Cycle

The reviewer should be aware that a normal EIS for a major federal action involves the following steps:

1. Publish notice of intent (NOI).
2. Hold public scoping meetings.
3. Prepare EIS implementation plan (IP).
4. Prepare working copy of draft EIS.
5. Lead agency internal reviews of working copy of draft EIS (DEIS).
6. Publish notice of availability (NOA) of DEIS in *Federal Register* (FR).
7. Public comments and public hearing(s).
8. Prepare review draft final EIS.
9. Lead agency internal review of review draft FEIS.
10. Publish NOA for FEIS in FR.
11. Publish record of decision (ROD) in FR.

Steps 4, 5, 8, and 9 are not strictly part of NEPA. But they are a necessary part of the NEPA compliance process because, during these internal reviews by senior level lead agency management and appropriate contractors, the full force of NEPA will be felt, and an ac-

Table 15
CONTENTS OF AN EA OR EIS

Foreword (Usually 1 to 2 pages)

1.0 Executive Summary (15 pages)
- 1.1 Purpose and Need
- 1.2 Proposed Action and Alternatives
- 1.3 Overall Evaluation of Alternatives
- 1.4 Mitigation Measures
- 1.5 Affected Environment
- 1.6 Environmental Consequences

2.0 Purpose and Need (1 to 2 pages)

3.0 Proposed Action and Alternatives (50 pages)
- 3.1 Proposed Action (25 pages)
  - 3.1.1 Background
  - 3.1.2 Relationship to Other Existing Facilities, if any
  - 3.1.3 Facility Description
  - 3.1.4 Process Description
  - 3.1.5 Process Effluents
  - 3.1.6 Resource Requirements
  - 3.1.7 Safeguards and Security
  - 3.1.8 Resistance to Normal Forces
  - 3.1.9 Reasonably Foreseeable Environmental Effects
- 3.2 Alternatives to the Proposed Action
  - 3.2.1 Alternative A
  - 3.2.2 Alternative B (etc.)
  - 3.2.3 Alternatives Considered but not Analyzed in Detail (optional)
- 3.3 Overall Evaluation of the Proposed Action and Alternatives

4.0 Affected Environment
- 4.1 Site Location and Regional Population
- 4.2 Regional and Site Activities
- 4.3 Socioeconomics
- 4.4 Physical Environment
  - 4.4.1 Geology-Topography
  - 4.4.2 Hydrology
  - 4.4.3 Seismicity
  - 4.4.4 Climatology
  - 4.4.5 Background Radiation
- 4.5 Ecology

(If more than one site is being considered, the affected environment at each site should be described.)

5.0 Environmental Consequences
- 5.1 Environmental Consequences of the Proposed Action
  - 5.1.1 Radiological Impacts (if nuclear facility)
  - 5.1.2 Nonradiological Impacts
  - 5.1.3 Impacts from Potential Accidents (very important for nuclear facilities)
  - 5.1.4 Security and Safeguards (important for nuclear and toxic chemical facilities)
  - 5.1.5 Unavoidable Adverse Impacts
- 5.2 Environmental Consequences of the Alternatives to the Proposed Action
- 5.3 Socioeconomic Effects
  - 5.3.1 Socioeconomic Effects of the Proposed Action
  - 5.3.2 Socioeconomic Effects of the Alternatives to the Proposed Action
- 5.4 Cumulative Effects
  - 5.4.1 Description of Nearby Facilities
  - 5.4.2 Cumulative Effects
- 5.5 Decontamination and Decommissioning

Table 15 (continued)
CONTENTS OF AN EA OR EIS

5.6 Relationship of Proposed Action to Land-Use Plans, Policies, and Controls
5.7 Irreversible and Irretrievable Commitments of Resources
5.8 Mitigation Measures

6.0 Pertinent Laws, Regulations, and Guidelines (list only really pertinent ones, the following are usually applicable to most industrial activities)
6.1 40 CFR 50, National Primary and Secondary Ambient Air Quality Standards
6.2 40 CFR 52, Prevention of Significant Deterioration of Air Quality
6.3 40 CFR 61, National Emission Standards for Hazardous Air Pollutants
6.4 40 CFR 141, National Interim Primary Drinking Water Regulations
6.5 40 CFR 260—266, Resource Conservation and Recovery Act

7.0 List of Preparers and Reviewers

8.0 Glossary

Appendix

References

Index

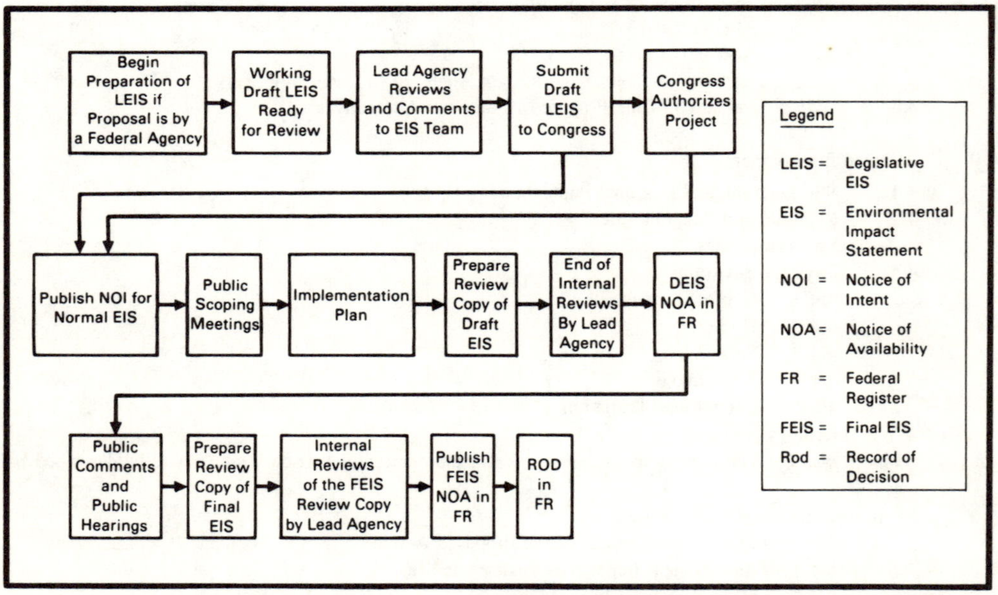

FIGURE 4. Steps in LEIS and normal EIS

ceptable document will be developed leading to an environmentally acceptable plant design. The time required to complete these reviews and obtain concurrence among all involved parties is a major factor that determines the NEPA compliance schedule. These reviews by senior level management and appropriate contractors in the NEPA compliance process have been found to be effective and beneficial in many projects. It was also found that full NEPA compliance allows public participation and provides the public a focal point for their criticism, and thereby perhaps avoids costly litigation and delays. Figure 4 provides a flow diagram of the NEPA compliance steps and review blocks.

Table 16
CHECKLIST FOR REVIEW OF EIS

1. Checklist for readability and clarity
   (a) Plain language, minimum technical jargon, all technical terms well defined
   (b) No ambiguous statements, tables, charts, graphs used to best advantage, no grammatical mistakes
   (c) Page limits set by CEQ generally met
   (d) Conflicting statements eliminated
2. Checklist for advocacy and bias
   (a) Conclusions obviously in conflict with data
   (b) Misinterpretation of conclusions
   (c) Conclusions not supported by data
   (d) Alternatives not treated equally
3. Checklist for technical quality
   (a) Use of most recently available, well-supported, well-referenced data
   (b) Use of approved models for calculating environmental consequences
   (c) Use of field-collected data by approved methods
   (d) Reasons for unreferenced data
   (e) Reasons for not providing needed data
   (f) Identified the unavailable but important data
   (g) Avoiding use of qualitative adjectives
   (h) All analytical procedures well described and meet professional standards
   (i) Controversial issues discussed
   (j) Environmental consequences presented for highest operating rates (bounding)
   (k) Address all applicable environmental laws at federal, state, and local levels
4. Checklist for environmental impacts presentation
   (a) Impacts presented in a balanced manner for all alternatives?
   (b) Impacts presented in table or charts for easy understanding by the public and the decision maker
   (c) Impacts presented in quantitative terms and compared with standards
   (d) Impacts presented so the significance can be understood by the public
   (e) Requirements of 40 CFR 1502.16 met
5. Checklist of mitigation measures
   (a) Best available control technology used; if not, reasons given
   (b) Mitigation steps clearly identified
6. Checklist for compliance with agency policy requirements
   The legal staff and senior management need to review the draft to make sure that commitments made in the EIS can be met and are in accordance with the lead agency policies.

Review of EIS by plant designers and engineers will help them understand the importance of designing the facility to reduce environmental insults. Also, commitments may have been made in the EIS to adopt mitigation steps that may or may not be possible of adoption. This type of error can be detected early in the planning stage if careful review is conducted by responsible supervisors in plant design, plant engineering, and policy/legal offices. Table 16 provides a checklist to aid the reviewers in six areas:

1. Readability, clarity, and understandability of EIS
2. Advocacy and bias
3. Technical quality
4. Environmental impacts presentation
5. Mitigation measures
6. Compliance with agency policy requirements

B. External Review Cycle

The lead agency is required to invite public comments on the draft and final EISs in accordance with 40 CFR 1503.1. The comments are to be invited from

1. Any federal agency which has jurisdiction by law or special expertise with respect to any environmental impact involved or which is authorized to develop and enforce environmental standards. Note that this includes the U.S. EPA.
2. Appropriate state and local agencies which are authorized to develop and enforce environmental standards.
3. Indian tribes, when the effects may be on a reservation.
4. Any agency which has requested that it receive statements on actions of the kind proposed.
5. The applicant.
6. The public, affirmatively soliciting comments from those persons or organizations who may be interested or affected.

Comments are to be invited on the final EIS before the decision (ROD) is made.

VIII. EPA REVIEW AND RATING OF EIS

The role of the U.S. EPA in the approval of the adequacy of the draft EIS is a major factor that determines whether the EIS satisfies the requirements of applicable federal environmental regulations. Under Section 309 of the Clean Air Act (42 U.S.C. 7609), the administrator of the Environmental Protection Agency is directed to review and comment publicly on the environmental impacts of federal activities, including actions for which environmental impact statements are prepared. If after this review the administrator determines that the matter is ''unsatisfactory from the standpoint of public health or welfare or environmental quality,'' Section 309 directs that the matter be referred to the Council on Environmental Quality (see also 40 CFR 1504.1). However, before the EPA Administrator refers the matter to CEQ, the EPA will try to work with the lead agency to identify mitigation methods or otherwise revise the proposed action, if possible, so that the environmental impacts can be eliminated or reduced to the point that the DEIS can receive a ''lack of objections'' rating.

The EPA has developed policy and procedures for review of environmental impact statements and has published a rating system as follows:

Rating of the impacts of the action			Rating of the adequacy of the impact statement		
LO	=	Lack of objections	Category 1	=	Adequate EIS
EC	=	Environmental concerns	Category 2	=	Insufficient information in the EIS
EO	=	Environmental objections	Category 3	=	Inadequate EIS
EU	=	Environmentally unsatisfactory			

Accordingly, LO is the highest rating available both from the view point of (1) environmental impacts of the action and (2) adequacy of the impact statement. Details of the EPA rating system are given in the attached summary issued by the EPA. EPA stated that ''the LO rating... has no numeric rating.'' Numeric ratings are associated with the following letter ratings below LO: EC (environmental concerns), EO (environmental objections), and EU (environmental unsatisfactory). More details of the EPA rating system are presented in Table 17.

IX. RESPONDING TO PUBLIC COMMENTS

After the notice of availability (NOA) of the DEIS is published by the lead agency in the

Table 17
SUMMARY OF THE EPA RATING, DEFINITIONS, AND FOLLOW-UP ACTION FOR DRAFT ENVIRONMENTAL IMPACT STATEMENTS (FROM EPA MANUAL 1640 POLICY AND PROCEDURES FOR THE REVIEW OF FEDERAL ACTIONS IMPACTING THE ENVIRONMENT)

Environmental Impact of the Action

LO — Lack of objections

The EPA review had not identified any potential environmental impacts requiring substantive changes to the proposal. The review may have disclosed opportunities with no more than minor changes to the proposal.

EC — Environmental concerns

The EPA review had identified environmental impacts that should be avoided in order to provide adequate protection for the environment. Corrective measures may require substantial changes to the preferred alternative or consideration of some other project alternative (including the no-action alternative or a new alternative). EPA intends to work with the lead agency to reduce these impacts.

EO — Environmental objections

The EPA review has identified significant environmental impacts that must be avoided in order to provide adequate protection for the environment. Corrective measures may require substantial changes to the preferred alternative or consideration of some other project alternative (including the no-action alternative or a new alternative). EPA intends to work with the lead agency to reduce these impacts.

EU — Environmentally unsatisfactory

The EPA review has identified adverse environmental impacts that are of sufficient magnitude that they are unsatisfactory from the standpoint of public health or welfare or environmental quality. EPA intends to work with the lead agency to reduce these impacts. If the potential unsatisfactory impacts are not corrected at the final EIS stage, this proposal will be recommended for referral to the CEQ.

Adequacy of the Impact Statement

Category I — Adequate

EPA believes the draft EIS adequately sets forth the environmental impact(s) of the preferred alternative and those of the alternatives reasonably available to the project or action. No further analysis or data collection is necessary, but the reviewer may suggest the addition of clarifying language or information.

Category II — Insufficient information

The draft EIS does not contain sufficient information for EPA to fully assess environmental impacts that should be avoided in order to fully protect the environment, or the EPA reviewer has identified new reasonable available alternatives that are within the spectrum of alternatives analyzed in the draft EIS, which could reduce the environmental impacts of the action. The identified additional information, data, analyses, or discussion should be included in the final EIS.

Category III — Inadequate

EPA does not believe that the draft EIS adequately assesses potentially significant environmental impacts of the action, or the EPA reviewer has identified new, reasonable available alternatives that are outside of the spectrum of alternatives analyzed in the draft EIS, which should be analyzed in order to reduce the potentially significant environmental impacts. EPA believes that the identified additional information, data, analyses, of discussions are of such a magnitude that they should have full public review at a draft stage. EPA does not believe that the draft EIS is adequate for the purposes of the NEPA and/or Section 309 review, and thus should be formally revised and made available for public comment in a supplemental or revised draft EIS. On the basis of the potential significant impacts involved, this proposal could be a candidate for referral to the CEQ.

Federal Register, the interested public usually submits written comments (and oral comments if a public hearing is held) on the DEIS to the responsible official in the lead agency. By law, a minimum of 45 days should be allowed for the public review and comment. The public here is defined as not only the private citizens, but interested environmental groups, and local, state, and federal agencies who have responsibility for protecting the environment.

It is not unusual to receive about 100 comments on a DEIS of medium complexity. On very controversial subjects, over 5000 comments can be received. The lead agency has to consider each comment very carefully. This can usually be delegated to the EIS contractor who screens the comments and logs them in a DEIS Public Comment Log of the type

Table 18
EXAMPLE FORM FOR LOGGING DEIS PUBLIC COMMENTS

DEIS Public Comments Log

Comment letter No.____________________ Date____________________

Commenting organization
or individual's name____________________________

1. Comment number	2. Full text of comment letter	3. Topic	4. Applicable EIS chapter/appendix	5. Where responded chapter/page/¶	6. Responding author	7. Cross reference to other comments

presented in Table 18. The log helps in making sure that each comment receives very careful analysis and appropriate response. The EIS project manager will then identify the DEIS author who can respond to the comment. The author can usually respond readily to a straightforward technical comment. But most comments are not that clear cut because the commentor can be emotional, and/or comment on issues that do not directly fall within the scope of the EIS. In such cases, the author selected to respond to the comment has to use judgment and respond accordingly. Table 19 provides one of at least ten ways an author can respond to a comment. There may be other types of responses possible.

When an author is certain of the response to a comment, he can complete the public comment-response form shown in Table 20. The EIS project manager and editor review the response and after the lead agency concurs with the response, the editor will incorporate it into the final EIS. The lead agency concurrence is necessary because the EIS is being written for the lead agency (who is the legal author) by the EIS contractor, if there is one. Sometimes the lead agency itself writes the EIS, in which case an EIS contractor would not be involved.

X. THE TIMING OF AGENCY DECISIONS

According to CEQ regulations (40 CFR 1506.10), no decision on the proposed action

Table 19
EXAMPLE ACTIONS FOR USE IN RESPONDING TO PUBLIC COMMENTS

Responding Author can Choose One of the Following Actions on the Public Comments

1. Comment addresses issue outside of scope of this EIS. Refer to lead agency.
2. Additional information requested is excessive. Need management decision.
3. Policy related comment. EIS project manager sends to lead agency for resolution and guidance before responding further.
4. Assumptions made by commentor invalid (provide explanation)____
5. Comparison of data/facts by commentor incorrect (provide explanation)____
6. Based on analysis, suggested data does not result in greater impacts than already reported in EIS (provide explanation)____
7. Agree that text/data is incomplete. Will perform additional analysis to complete (provide explanation and revised text/data)____
8. Editorial comment. Prefer current wording. (Provide explanation)
9. Editorial comment. Will redo. (Provide revised text)
10. I have verified that data are unavailable at this time, but ROD program is expected to provide the data at a later date. (Provide explanation)

which is the subject of an EIS should be made or recorded until after the following dates, whichever is later:

1. 90 days after publication of the notice of availability of DEIS
2. 30 days after publication of the notice of availability of FEIS

Also, not less than 45 days shall be allowed for comments on DEIS. There is no requirement for the agency to request comments on the FEIS, but the requirement to wait 30 days after publication of the FEIS to publish a final decision is intended to ensure that the decision maker has time to consider the information in the EIS prior to final decision making, and perhaps to allow the public to react to the FEIS in those 30 days. If the public still wants to challenge an agency decision after the 30-day waiting period and publication of the ROD by the agency, resorting to the courts is an available method of challenge. However, when an agency has complied with the law by strictly enforcing the NEPA requirements in letter and spirit, the courts may take a dim view of the public suit against the agency decision.

XI. SCHEDULING AND COSTING THE NEPA PROCESS

On any given project, the duration of the work for the NEPA staff assigned to that project is a function of the complexity, the significance, the controversy, and the severity of the

Table 20
EXAMPLE OF PUBLIC COMMENTS-RESPONSE FORM

Commenting agency/individual ____________________

Comment number (from log) ____________ Cross reference to other comments ____________

Responding author ____________ Date in ____________ Date out ____________

Response reviewed by ____________ (EIS project manager) ____________ (Editor) ____________ (Lead agency)

Response incorporated in EIS on ____________ (Date) by ____________ (Editor)

Author's Response

Author suggests to include response on EIS page ____________, line ____________.

Note to Author: In preparing your response, please be aware that there are perhaps similar comments from other commentors and that these are suitably considered.

expected impacts of the proposed action. Another important item that will determine the duration of an EIS project is the amount of available background information necessary to do the impact analysis. If sufficient background data are not available for the impact analysis, the data would need to be generated (unless valid reasons can be given why generating the data will be resorted to). To summarize, the duration of the NEPA compliance process depends on the following factors:

- Project size and type (nuclear or nonnuclear is a good example)
- Project controversy and degree of public concern
- Significance of estimated impacts
- Availability of background information
- The location of the proposed action (local attitudes, class of the area relative to PSD permit, etc.)
- The time allotted for reviews

There have been projects which have taken more than 5 years for the NEPA compliance process completion. However, as a general rule, if the NEPA process definition is limited to the effort involved between publishing a notice of intent (NOI) and a record of decision (ROD), the following approximate durations apply:

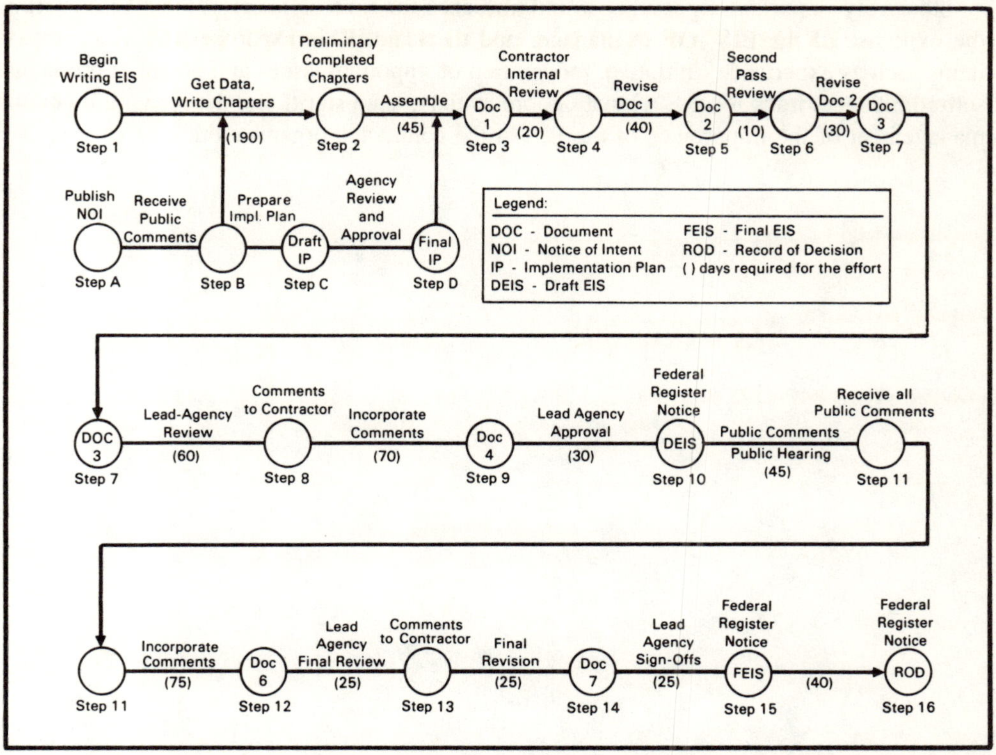

FIGURE 5. Generalized logic and time flow diagram for EIS process

1. 18 months for projects of low complexity, low severity of impacts, and minor controversy of projected impacts.
2. 24 months for projects of medium complexity, severity, and controversy of projected impacts.
3. 36 months (or more) for projects of major complexity, severity of impacts, and controversy of projected impacts.

These durations are to be used for rule-of-thumb estimates and are not set in concrete. Figure 5 is intended to provide the reader with a format useful in estimating the time needed to complete the NEPA Process.

Example exercise — The reader is encouraged to estimate the duration of the EIS process for the coal-fired power plant construction and operation described earlier.

Cost of NEPA compliance — An EIS project can require anywhere from $10,000 to $10,000,000, depending on the same factors that influence the time required to complete the process. These factors that influence the cost are

- Project size and type
- Project controversy in the public eye
- Need for public hearings, workshops, and scoping meetings
- Severity and significance of environmental impacts
- Availability of background data
- The location of the action
- Time allotted for reviews

Another very important aspect that would influence the time and cost of the NEPA work is the expertise of the EIS project manager and the team. This expertise would encompass items such as experience, initiative, recognition of important issues and potential problems, self-confidence, track record, reputation, credentials, team spirit, and the recognition by the management of the importance of the EIS in the contracted organization.

Chapter 6

SELECTING CONTROL TECHNOLOGIES FOR COMPLIANCE WITH REGULATIONS

I. INTRODUCTION

We have seen in Chapter 3 (Technical Aspects of Statutes, Standards, and Regulations) and in Chapter 5 (How to Write Environmental Documents) that the NEPA process implementation often leads to a specific requirement for mitigating environmental impacts. Frequently, such mitigation is best built into the design of the proposed project using control technologies that either eliminate the emission of pollutant(s) to the environment or reduce them. Some of the methods of achieving this visible benefit of the NEPA process are described in this chapter. The approach to emission control methods (most preferred to not preferred) is shown in Figure 1.

II. AIR POLLUTION CONTROL TECHNOLOGIES

In a book of this type, the vast field of air pollution control technology selection can be barely touched upon. At best, the following discussion can be no more than a brief introduction to air pollution control technologies. Anything more detailed is beyond the scope of this book.

Compliance with the air pollution control regulations and emissions standards promulgated by the federal and state agencies often requires a substantial reduction, or occasionally elimination, of emissions from a source. This can be achieved in one of two ways:

1. Prevent pollution at the source by substituting the raw material ingredient that causes the pollution or by process change.
2. Control pollution at the source after the fact by using add-on control devices to capture the pollutants.

A. Preventing Pollution at the Source

To start off with an example, the 3M Company considers preventing pollution at the source to be more efficient and less costly than after-the-fact, end-of-the-pipe treatment or capture.[18] Koenigsberger describes four basic ways to prevent pollution at the source:

1. Develop nonpolluting products or reformulate existing products by substituting non-polluting materials for ingredients that are pollutants. A good example of this is the effort of 3M to substitute water-based adhesives for those that contain solvents. The untreated solvents are pollutants that contribute to the formation of ozone in the atmosphere.
2. Modify production operations-conversion to solventless coating processes or changing from a batch to a continuous process to reduce peak discharges and allow for water recycle.
3. Redesign equipment used in a manufacturing process. A plant boiler was converted to burn solvent-laden exhaust from a maker, thereby eliminating add-on pollution control equipment, resulting in energy savings of up to $270,000 per year.
4. Recover and recycle waste products. As an example, at one of the plants, a film developing unit generated wastewater containing 1,1,1-trichloroethane developer. A decanter system was installed to recover the developer for later distillation and recycle.

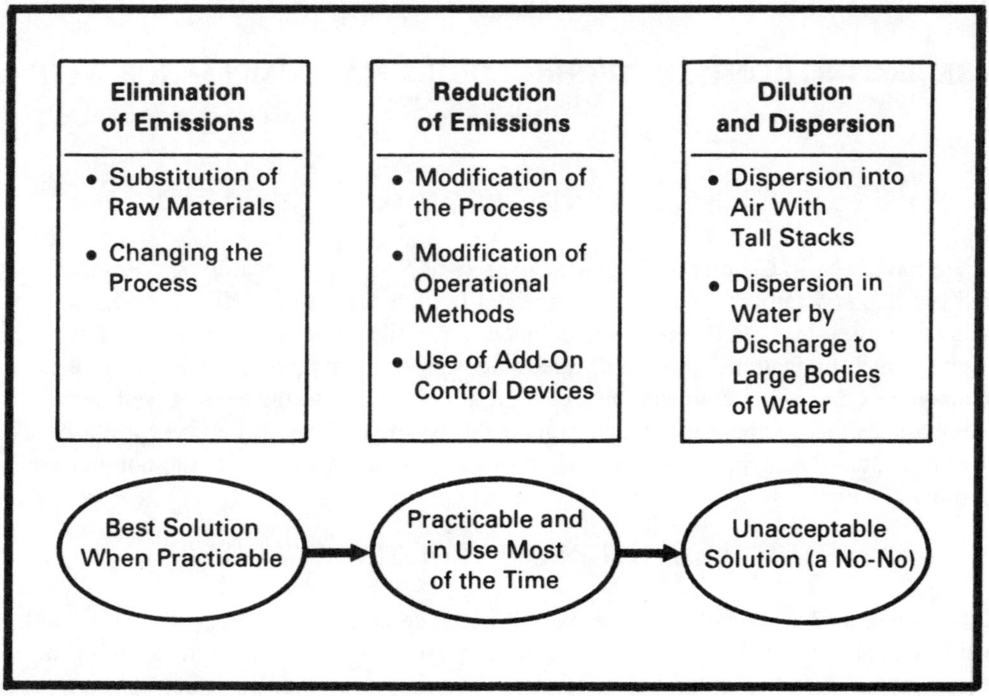

FIGURE 1. Approach to emission control methods.

> The installation cost of the decanter system was \$4000 and it saved over \$20,000 during the first 4 years of operation.

These are examples of efforts to eliminate air pollution emissions before the pollutant enters the plant effluent stream. Stern[31] also gives examples of how air pollution emissions can be eliminated under two categories: (1) substituting products and (2) changing processes.

1. By Substituting Products

Examples in industry in which air pollution was entirely eliminated by product substitition are (1) substitution of rubber with a cold-setting synthetic resin in the manufacture of paint brushes, thus eliminating the need to vulcanize the rubber bond at the base of the bristles and the resultant severe odor nuisance during vulcanization, (2) elimination of the use of lead "anti-knock" compounds to improve octane number of gasolines used in motor vehicle fuel, and (3) substituting low-sulfur coal to replace the use of high-sulfur coal in coal-fired electric plants, thereby avoiding high sulfur dioxide emissions.

2. By Changing Processes

Changing a process to reduce or eliminate air pollution has been successfully used in many industries including coal-fired power generation, petroleum refining, and agriculture. In power generation, using low excess air in the combustion zone has not only reduced the formation of sulfuric acid in the combustion gases, but also reduced the formation of nitrogen oxides when secondary air is injected at lower temperature zones. In petroleum refining, recycling of vapors to the process by condensation of low boiling components (from petroleum storage tanks, for example) has not only reduced product loss, but eliminated air pollution. In agriculture, the use of liquid fertilizers applied to the earth by injection has reduced the fugitive dust pollution which is common when fertilizer is applied as dry powder. Also, loss of the fertilizer to wind is eliminated.

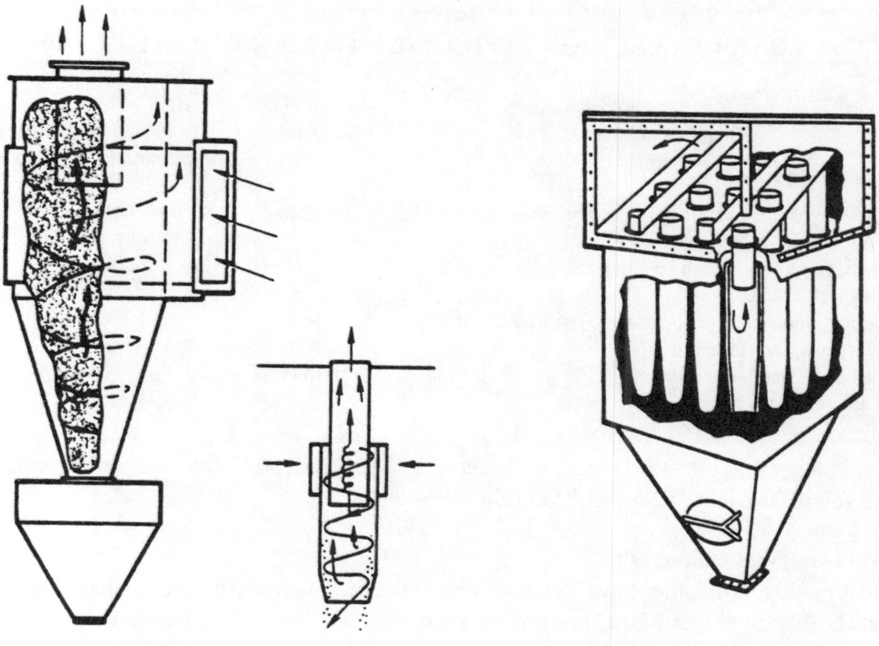

FIGURE 2. Examples of dry collection devices.[21]

B. Controlling Pollution by Using Add-On Devices

More often than not, add-on controls are used to effectively reduce (not eliminate) pollution. Add-on control devices may be defined as devices placed in the effluent stream such that the pollutant-bearing effluent enters and exits the device after depositing most of the pollutant in the device. Add-on controls are usually the only method of pollution control available for plants that were operational before control requirements became law. Numerous gas cleaning devices used in electricity generation, heating plants, Portland cement plants and other industries wherever dust emissions need capture are add-on types as follows:

- Mechanical collectors (cyclones, etc.)
- Scrubbers (wet collectors)
- Electrostatic precipitators
- Fabric filters
- High efficiency particulate air (HEPA) filters

1. Mechanical Collectors

Also called inertial devices, mechanical collectors use centrifugal force to drive the dust particles in a gas stream toward the wall of a collector such as a cyclone by a radial force (see Figure 2). Near the wall, a concentrated dust layer is formed. This layer slowly swirls down the wall of the cyclone and deposits the dust into the discharge area located near the bottom. Reentrainment of the dust into the main gas stream usually occurs and reduces the collection efficiency. As a class of equipment, cyclones offer the lowest efficiency of collection as well as lowest cost for installation and operation. Cyclones are usually used as first-stage collection devices to remove coarse particles, since even a high-efficiency cyclone has poor efficiency for particles below 5 μm, the size of long-damaging particles that stay suspended in ambient air (Table 1).

Table 1
CYCLONE COLLECTION EFFICIENCY FOR SOME PARTICLES[19,31]

Typical particle description	Size range (μm)	High-efficiency cyclone collection efficiency (approximate wt % collected)
Fume, soil smoke, tobacco smoke, carbon black, zinc oxide fume	Less than 1	Almost 0
Lung-damaging dust, fly ash components, sulfuric mist, insecticide dust, bacteria	1 to 5	60
Cement dust, most of fly ash particles, fertilizer, bacteria	5 to 20	85
Coarser fly ash, human hair	20 to 40	97
Beach sand, ground lime stone, coarse fly ash, drizzle, human hair	Above 40	98

Detailed information on selection, design, and specifications for mechanical collectors can be found in Licht,[19] Moore,[21] and Stern.[31]

2. Scrubbers (Wet Collectors)

Wet-type dust collectors have the dual advantage of (1) removing the contaminants more efficiently than cyclones and (2) removing both inert dust particles as well as toxic gaseous components from the dirty air. Various types of wet collectors in use include packed columns, plate columns, spray towers, wet centrifugal towers, and venturi scrubbers.

Unlike dry dust collection devices, a wet scrubber can remove gases as well as particulates. The high dust collection efficiency of scrubbers is proportional to the electrical power input related to the pressure drop of the gas and/or liquid across the venturi throat (see Figure 3), hence the operating cost. Further, the collected dust is wet; handling wet dust and sludge is more difficult than handling dry dust. Since an electrostatic precipitator can gather dry dust, the dilemma for a new installation, such as the coal-fired power plant that was described in Chapter 5, would be whether to collect most of the dust dry with an electrostatic precipitator followed by the SO_2 removal with a wet scrubber or employ a scrubber to do both the jobs concurrently. The choice is difficult. In geographic areas like Arizona where rainfall is small and the weather mostly dry, land disposal of wet sludges is usually feasible. Thus, the choice is also site-situation related.

Scrubbers occupy a moderate amount of plant space compared to, say, electrostatic precipitators. But sludge ponds will be needed. Capital costs of scrubbers are lower. The cleaned gas is cool and wet; reheating may be necessary for restoring plume buoyance. Scrubber maintenance can be tedious and expensive, particularly if the system has a tendency to deposit scales.

More than 100 different scrubber designs are available and more are evolving. Although differing in specific constructional and hardware component arrangement details, all these scrubber designs employ the basic eight types, as listed in Table 2. Each of these scrubber types has been applied to remove SO_2 from coal-fired boiler flue gases. Their other applications are listed also in Table 2. Table 3 is a partial list of scrubber suppliers.

Collection efficiency of scrubbers can be predicted theoretically. But, practically speaking, wet scrubbers design and selection is almost always based on actual performance tests. Obviously, the cleaning efficiency of scrubbers is proportional to the energy input, i.e., pressure drop and the liquid-to-gas ratio expressed as gallons per 1000 actual cubic feet per minute (ACFM) of gas.

In general, venturi scrubbers offer very high collection efficiency (>99% for particulates 1 μm or larger and 90 to 99% for submicron particulates). Pressure drops over the throat range from 5 to 60 in. of water gauge. Liquid-to-gas (L/G) ratio ranges from 10 to 25 gal/1000 ACFM.

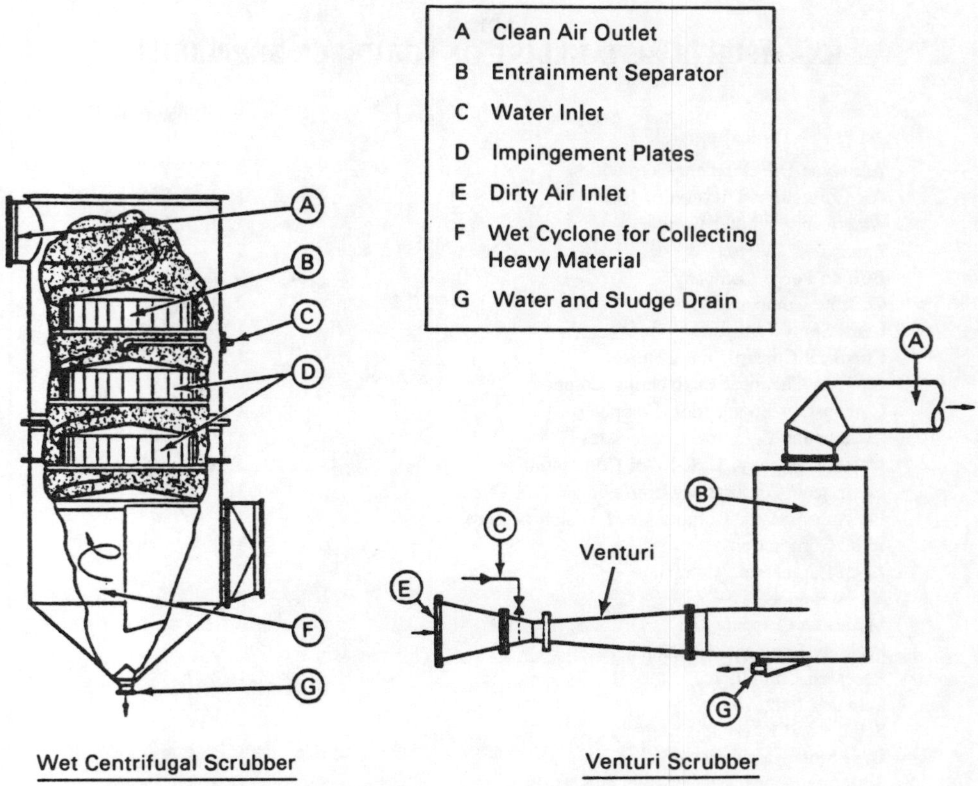

FIGURE 3. Examples of wet-type dust collectors.[21]

Table 2
BASIC SCRUBBER TYPES AND PARTIAL LIST OF APPLICATIONS

Scrubber type	Scrubber type no.	Some applications[a]
Venturi scrubbers	1	Used in almost all dust and gaseous pollutant removal in nearly every industry
Packed (plate and floating bed) towers	2	Acid mist and soluble dust removal, floating bed for fly ash and gummy precipitate removal
Impingement scrubbers	3	Fly ash, oil mist, coal dryer dust, blast furnace, sintering machines, calciners, etc.
Mechanically induced spray scrubbers	4	Blast furnace flue gas, steel industry foundaries, coal processing, etc.
Spray towers and spray chambers	5	Soot blowing operations, cooling and humidification, boiler stacks
Cyclonic scrubbers	6	Dust removal from lime kilns, reverberatory and blast furnaces, and rotary driers
Centrifugal scrubbers	7	Same as cyclonic scrubber
Irrigated filters and mist eliminators and wire mesh tray scrubbers	8	Mist removal in scrubbing towers, mist removal from sulfuric, nitric, and phosphoric acid plant, and oil mist removal

[a] All these types have been employed in pilot/demonstration SO_2 scrubbing with or without modifications.

From Engdahl, R. B. and Murthy, K. S., Fourth Int. Pollution Control Congr., Cleveland, Ohio, October 20 to 23, 1975.

Table 3
EXAMPLE (PARTIAL) LIST OF SCRUBBER SUPPLIERS

Supplier	Scrubber types[a]
Aerodyne Development Corporation	1, 2, 5, 7, 8
Air Corrections Division of UOP	1, 2, 5, 7, 8
American Air Filter Company	1, 3, 4, 5, 7
Babcock & Wilcox	1
Buffalo Forge Company	5, 7
Ceilcote Company, The	2, 3, 5, 7
CentriSpray Company	1, 3, 4, 5, 7
Chemical Construction Company	1, 2, 5, 7, 8
Chiyoda Chemical Engineering Company	7
Combustion Engineering Company	1
Combustion Equipment Associates	1, 3, 6
Ducon Company, U.S. Filter Corporation	1, 5
Envirotech Corporation, Emission Control Division	1, 3, 5, 7
Fly Ash Arrestor Corporation, Division of Zurn	1
FMC Corporation	1, 2, 5
Koch Engineering Company	2
Krebs Engineers	5
Monsanto Company	8
Peabody Engineering Company	1, 3, 5, 7
Research-Cottrell, Inc.	1, 2, 7
Schmieg Industries	4, 5, 7
Schutte and Koerting Company	1
U.S. Stoneware Company	1, 2, 5, 8
Sly Manufacturing Company, The W. W.	1, 3, 7
Western Precipitation Division, Joy Manufacturing Company	1, 3, 4, 5, 7
York Company	8
Zurn Industries, Inc.	1, 3, 6, 7

[a] The numbers 1 through 8 refer to Scrubber type no. listed in Table 2.

From Engdahl, R. B. and Murthy, K. S., Fourth Int. Pollution Control Congr., Cleveland, Ohio, October 20 to 23, 1975.

Packed plate and floating bed scrubbers usually have less than 99% efficiency. Pressure drops vary from 2 to 4 in. H_2O per stage, and usually about four stages are used. Liquid-to-gas ratio changes from 5 to 20 gal/1000 ACFM. The turbulent contact absorber (TCA), a UOP design, is an example of floating bed scrubbers.

Impingement scrubbers collect particulates with somewhat lesser efficiency than venturi scrubbers, but the pressure drops (2 to 30 in. H_2O and L/G ratios (2 to 5 gal/1000 ACFM) are both lower. Thus, it is helpful to evaluate impingement scrubbers along with others during any specific application.

Mechanically induced-type wet scrubbers have a much lower collection efficiency (90 to 95% on 1 μm or larger, and 99% plus on 5 μm or larger). Pressure drops range 3 to 10 in. H_2O, and L/G ratios are 0.5 to 5 gal/1000 ACFM.

Spray towers and chambers usually have the lowest dust collection efficiency (80% for particles 10 μm or larger). However, the pressure drops are very low, 0.5 to 1.0 in. H_2O, and L/G ratios are 5 to 10 gal/1000 ACFM. It is possible to improve spray-type scrubber efficiencies by modifications.

Pollution control by using add-on devices has created a multibillion-dollar equipment vending and installation industry, as evidenced by numerous companies who supply scrubbers (Table 3).

3. Flue Gas Desulfurization Technology

Flue gas desulfurization (FGD) is a widely used method in coal-fired power plants to meet the SO_2 emission standards which are designed to hold the ambient SO_2 levels within the ambient air quality standards. The impact of just this one standard for SO_2 emission control on the industrialization of the traditional coal-fired power plant by converting it to a sophisticated chemical plant facility can be visualized in Figure 22. While the scheme described in Figure 4 is not exactly in use in all coal-fired power plants, many facilities use FGD systems that look somewhat like it. Most such facilities do not produce sulfur as the end product of FGD. But if sufficient sludge disposal and management facilities are unavailable, FGD systems that do not produce sludge could become a necessity.

In the likely new look of coal-fired power plants shown in Figure 4, since sulfur is the desired product, the system describes a process of converting the SO_2 to S using the following reducing agents: hydrogen sulfide (H_2S), carbon (C), carbon monoxide (CO) and hydrogen (H_2), and methane (CH_4). The reduction reactions are as follows:

- Claus process:

  $$2\ H_2S + SO_2 \rightarrow 1.5\ S_2 + 2\ H_2O$$

- RESOX process:

  $$C + SO_2 \rightarrow S + CO_2$$

- ASARCO process:

  $$CO + 2H_2 + 1.5\ SO_2 \rightarrow 0.75\ S_2 + CO_2 + 2\ H_2O$$

- Allied Chemical reduction process:

  $$CH_4 + 2\ SO_2 \rightarrow CO_2 + 2\ H_2O + 2\ S$$

According to Rubin et al.,[30] if a national acid rain control statute were enacted, the retrofitting with FGD systems to remove up to 90% of the sulfur dioxide emissions from existing coal-fired power plants (that are perhaps operational from before 1970) may increase significantly.[30] The FGD technology retrofit installation costs could be as high as $240 million for a 1000-MW electric plant. The operating cost could be as high as 11.0 mil/kWh (i.e., 1.1 cents per kWh) of energy generated.[30]

4. Fluidized-Bed Combustion Technology

For new coal-fired power plants to be built in the future, process change and process modification could be an available option to the use of FGD add-on systems. The newly developed fluidized-bed combustion (FBC) boilers have the potential to reduce not only SO_2, but also NO_x emissions, while improving boiler efficiency and eliminating the wet sludge generation. Fluidized-bed combustion is an available technology which should be considered as an alternative control technology.[23]

Fluidized-bed combustion of coal takes place at about 900°C which is significantly lower than the 1650°C employed in conventional coal-fired boiler systems. At the lower temperature, slag is not formed and volatilization of alkali salts is reduced. Of greater importance, formation of nitrogen oxides is significantly reduced and most of the SO_2 formed reacts with the bed material to form $CaSO_4$. This is withdrawn from the system with the ash. Thus, FBC produces both soft ash and low SO_2 and NO_x emissions, eliminating the need for add-

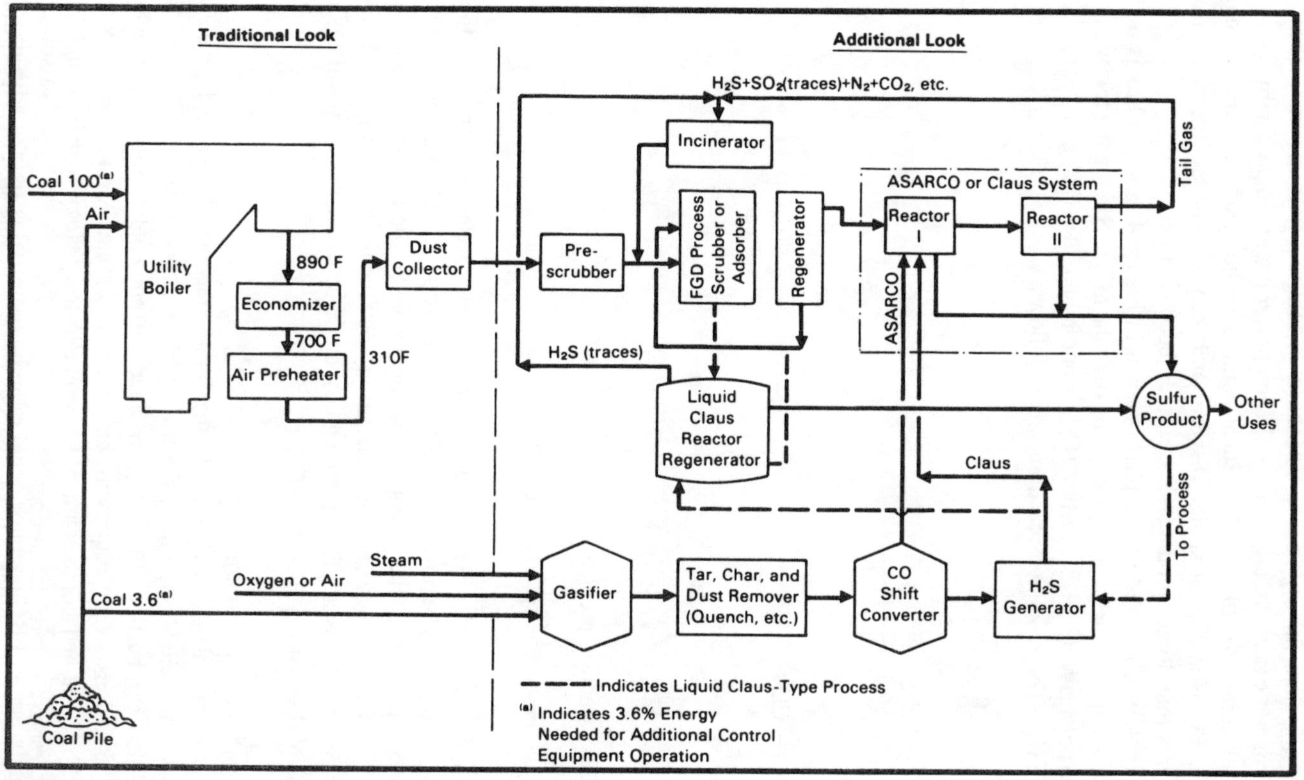

FIGURE 4. An example of the new look of coal-fired power plants. (From Murthy, K. S., Rosenberg, H. S., and Engdahl, R. B., *J. Air Pollut. Control Assoc.*, 26, 9, 1976. With permission.)

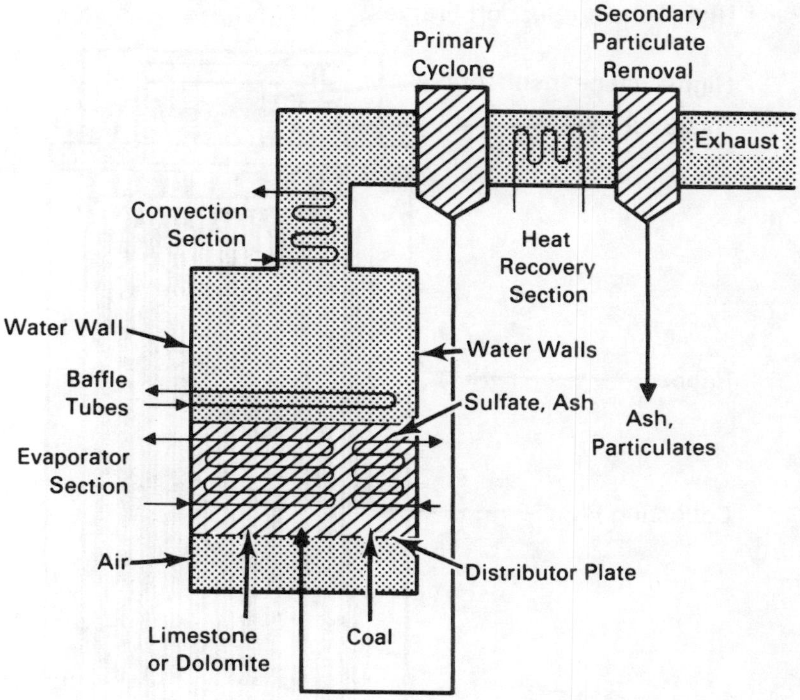

FIGURE 5. Fluidized-bed combustor schematic diagram.[23]

on or after-the-fact emission control devices. A schematic diagram of an FBC is shown in Figure 5.

Heat is continuously extracted from the FBC bed. This is accomplished either with immersion cooling tubes or by supplying excess combustion air to carry away the heat. Heat transfer between the bed and cooling tubes is both radiative and convective. Coefficients can be up to ten times larger than in conventional gas-to-surface heat exchange systems. Furthermore, in a conventional unit, only half of the heat exchange tube surface is exposed to combustion gases. The immersed tubes in FBC are totally exposed with a significant savings in tube surface area. Lower combustion temperatures also reduce corrosion and fouling of the heat exchange tubes. However, tube erosion may be a problem.

Higher heat transfer rate permits a dramatic decrease in boiler size, which can be further decreased by pressurizing the system. Atmospheric pressure FBC systems, which are currently receiving greater emphasis, transfer combustion heat to water circulating in the immersed tubes and convert it to steam. In a pressurized system, combustion can take place at pressures up to 10 atm. The hot gases leaving the bed not only make steam as do atmospheric pressure systems, but are also expanded through a gas turbine for electric power generation. For both systems, the reduced boiler size results in significant reduction in capital and maintenance costs.

FBC can be used efficiently with almost any fuel, including municipal wastes, paint residues, sewage sludge, and oily wastes. A multifuel system permits switching to solid, liquid, or gaseous fuels, so an FBC operator could quickly and efficiently take advantage of changes in fuel costs and availability.

Fluidized-bed combustors have the potential for increased coal-burning efficiency which can be as high as 45% in a pressurized FBC, compared to about 34% for conventional coal-fired units with add-on emission controls. Together, the increased efficiency and multifuel capability have the potential for substantial savings in the consumption of coal and oil.

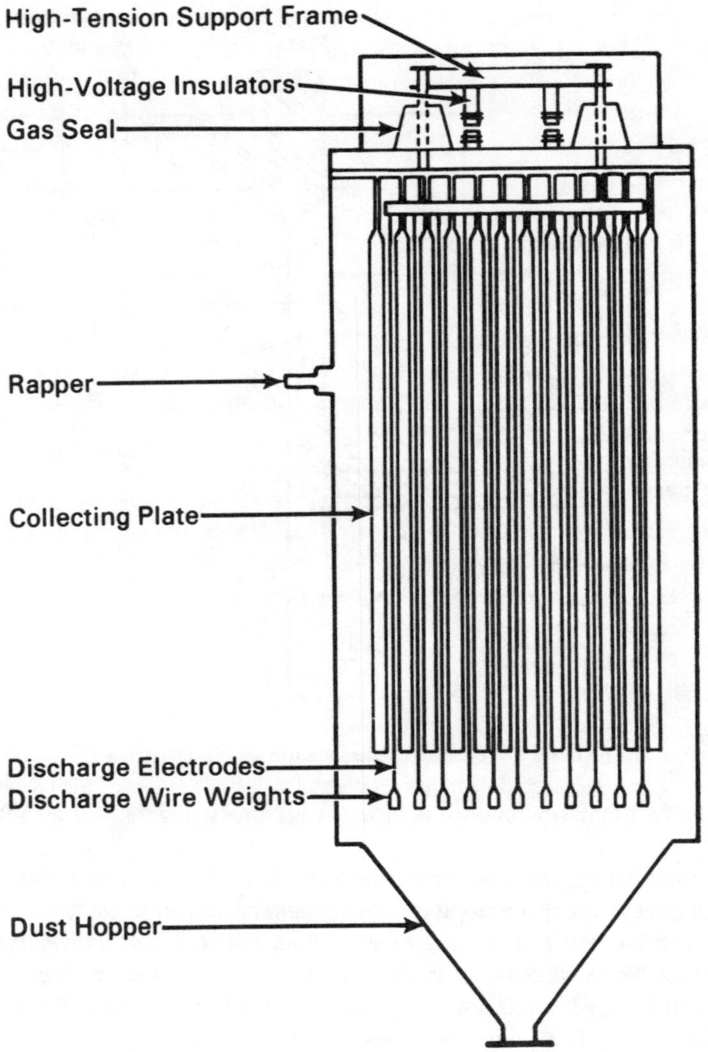

FIGURE 6. Electrostatic precipitator schematic diagram.[21]

5. Electrostatic Precipitators

Electrostatic precipitators (ESPs) are used as dust collection devices and to remove aerosol liquid particles (such as sulfuric acid mist) from air streams. ESPs have been successfully used in the metal smelter industry (copper smelting), lead blast furnaces, and in cement film dust control. From these beginnings, the use of ESPs expanded to the removal of fly ash from the flue gases generated in coal-fired boilers in electricity generation.

The collection principle of the ESP is very simple in concept: (1) The dust particles receive an electric charge discharged by an electrode, (2) the charged dust particles are attracted to oppositely charged collecting plates, and (3) the plates are periodically rapped to loosen the dust and remove it into a dust hopper. The structure of an ESP is schematically shown in Figure 6. A detailed description of the electrostatic precipitator including the methods of designing, selecting, and installing can be found in Stern.[31]

The collection efficiency of ESP is a function of the collection plate area available in the precipitator and the velocity of gas through the precipitator. When these are favorable, more than 99% of particles of 1 μm in size can be collected and nearly 100% of all particles over 3 μm can be removed.

The advantages of electrostatic precipitation are that, unlike in scrubbers, the gas need not be pressurized, the energy required for gas cleaning is less than that for equipment in which the entire gas stream is pressurized, and the operating costs are lower. The disadvantages of ESP are very high first costs (installation and equipment costs) and large space requirements for the higher efficiency ESPs.

6. Fabric Filters

Fabric filters, or bag houses, capture and retain particles on the fibers of the filter cloth. These filters are highly efficient particulate collectors, especially for submicron particles. The high efficiency of collection is obtained by the build-up of a mat or cake of the dust on the bag cloth, the cloth serving as a support structure. The thicker the cake, the higher the collection efficiency; however, the thicker cake also causes very high resistance to the gas flow. Therefore, the cake must be dislodged and collected periodically.

Fabric filters with a minimum guaranteed collection efficiency of 99.5% on submicron particles are available. Because of this reliability, they are extensively employed to control harmful or obnoxious emissions from operations involving abrasives, irritating chemical dusts, oil-fired boilers, and oxygen-fed convertors for steel manufacture.

The material for the bag can be selected from among cotton, wool, paper, nylon, orlon, glass, graphitized fiber, asbestos, etc. The selection of the material depends on the characteristics of the gas stream, i.e., if it is not acidic, basic, and has organic solvents.

7. High-Efficiency Particulate Air Filters

These filters, commonly called HEPA filters, are defined as a throwaway, dry-type filter having (1) a minimum particle removal efficiency of greater than 99.97% for 0.3 μm-sized particles, (2) a maximum resistance, when clean, of 250 Pa (1 in. H_2O) when operated at rated air flow capacity, and (3) enclosed in a rigid casing extending the full depth of the filter medium.[21]

High-efficiency particulate air filters find practical applications in the nuclear industry and medical laboratories where radioactive dust or bacterial particulates need to be effectively controlled. Clean rooms, electronic instrument assembly areas, operating rooms, etc. also use HEPA filters. High efficiency particular air filters are not used in large industries like metals, electric power generation, cement kilns, etc.

C. Example of Air Pollution Control Technology Application

Let us illustrate the application of air pollution control technology to the coal-fired power plant discussed earlier in Chapter 5. In this power plant, the two key pollutants that require control using add-on devices are SO_2 and TSP. Both these pollutants are present in the flue gas. The flue gas flow rate from each of the 500-MW units is calculated to be about 1.8 million actual cubic feet per minute (ACFM) at an average temperature of 250°F and a pressure of 1 atm before the flue gas enters the particulate control device. The dust loading in this gas is calculated to be 8.3 gr per standard cubic feet (SCF), or about 96,000 lb/hr. To meet the NSPS for particulate matter, the emissions will have to be reduced to less than 0.03 gr/SCF or about 99.7% reduction. To provide a margin of safety, let us design the control equipment to provide about 99.8% collection. This is quite a high efficiency of collection that would be required.

Both fabric filters and electrostatic precipitators are capable of providing the high efficiency of collection required in this example. But the operating cost of the control equipment will be increased disproportionately at efficiencies above 99.5%. The last 0.3 to 0.4% collection can increase not only the operating cost disproportionately, but also the equipment size and cost because efficiency is a function of collection area. Also, removal of SO_2 from the flue gas is another requirement imposed by the NSPS and PSD requirements. Calculations show that about 90% removal of SO_2 would be needed.

Both hot-side (located before the air preheater) and cold-side (located after the air preheater) precipitators are available with guaranteed removal efficiencies in excess of 99.5%. Cold-side precipitators typically operate at 250 to 350°F, while hot-side precipitators typically operate at 600 to 850°F. Because of the higher temperatures associated with hot-side precipitators, the volume of flue gas is higher, resulting in increased precipitator size and cost. The electrical resistivity of the fly ash, a function of the type of coal and the temperature of the flue gas, significantly affects performance of the precipitator. Thus, there is an optimum range of resistivity values for maximum collection efficiency. Inadequate collection efficiencies resulting from problems with fly ash resistivity can be revolved by increasing precipitator size, by adding chemical conditioning agents, or by not removing too much sulfur during coal cleaning. Typically, precipitators perform more efficiently using coals with high sulfur content, and this was one of the reasons for not electing the cleaned coal alternative in this example.

Fabric filters are also available with guaranteed removal efficiencies in excess of 99.5%. As particulate collection requirements become more stringent, it is anticipated that fabric filters will be more widely used and may offer economic advantages over electrostatic precipitators.

Wet scrubbers utilize an injected spray to make contact with the fly ash particulates in the flue gas stream. When the spray and particles collide, the particles adhere to the surface of the spray and are then removed from the gas stream. Particulate collection efficiency of a wet scrubber depends on the degree of atomization of the scrubber liquor, size distribution of the fly ash particles, and concentration of the particles in the flue gas stream. Depending on these factors, relatively high efficiencies can be attained by wet flue gas scrubbing. Presently, wet scrubbers have not proven as efficient in particulate collection as electrostatic precipitators, especially for submicron particles. Typically, wet scrubbers also remove 10 to 20% of the SO_2 in the flue gas stream as well as particulates.

The wet scrubber has a slight advantage over electrostatic precipitators with respect to the use of natural resources when used in combination with a lime/limestone scrubber (for SO_2 removal). The slightly alkaline fly ash entrained in the wet particulate scrubber may reduce limestone requirements, but this benefit would be partially offset by the higher energy consumption and maintenance requirements of a wet scrubber, compared to an electrostatic precipitator.

The installed cost of a high-efficiency wet particulate scrubber is comparable with that of an electrostatic precipitator of similar performance. However, the operating cost of the wet scrubber may be as much as ten times that of an electrostatic precipitator because of the very high pressure drop required for high-efficiency operation (up to 30 in. water gauge). Furthermore, the scrubber system requires additional sludge handling and disposal and has exhibited greater maintenance requirements. For these reasons, the wet particulare scrubber is not considered an optimum particulate removal system for this plant.

For this facility, an electrostatic precipitator to limit particulate emissions was selected according to an economic and engineering evaluation based on proposals received. Another reason for selection of the electrostatic precipitator is that the ESP, when used in combination with a limestone scrubber for SO_2 removal, can provide very high particulate control. Any particulates not collected in the ESP will be scrubbed out in the wet scrubber along with the SO_2. For the example case, Table 4 provides a comparison of the two particulate control systems considered.

On the face of it, the comparison shows an advantage for the bag house. The power consumption for the ESP is higher. Yet the maintenance problems for an ESP are much less, particularly, there are no bags needing to be replaced. This reason, coupled with many other considerations including the use of a SO_2 scrubbing unit, led to the selection of ESP over the bag house.

Table 4
COMPARISON OF PARTICULATE REMOVAL SYSTEMS FOR ONE 500-MW UNIT

	Electrostatic precipitator	Bag house
Estimated equipment cost	$15,000,000	$12,000,000
Design removal efficiency	99.83%	99.83%
Power consumption	2,150 KW	400 KW
Maximum pressure drop across the system	2.5 in. H_2O	6.0 in. H_2O
Total power cost including fan power cost per year	$11,000,000	$7,500,000
Labor and maintenance cost per year	$60,000	$280,000

Table 5
WATER POLLUTANT CATEGORIES AND SOURCES

Pollutant category	Possible sources
BOD/DO	Municipal wastewater treatment plants; industries (particularly pulp and paper mills); combined sewers; natural sources, in general, oxygen demanding wastes
Bacteria (disease-causing agents)	Municipal wastewater treatment plants; combined sewers; urban runoff; feed lots; pastures and rangeland; septic systems; natural sources
Nutrients	Municipal wastewater treatment plants; agriculture; septic systems; silviculture; combined sewers; construction runoff
Suspended solids/turbidity	Agriculture; urban runoff; silviculture; construction runoff; mining; industries; combined sewers
Total dissolved solids	Agriculture; mining; urban runoff; combined sewers
pH	Atmospheric deposition; mine drainage
Ammonia	Municipal wastewater treatment plants; combined sewers
Metals and toxic substances	Industries; municipal wastewater treatment plants; agriculture; land disposal of wastes; silviculture; urban runoff; spills; combined sewers
Synthetic organic compounds	Detergents and pesticides industry, one typical example, hazardous waste disposal
Oils	Petroleum industry; used oil disposal
Radioactive materials	Mining; processing uranium ore; milltailing disposal
Heat (thermal pollution)	Thermal discharges from electric power plants and other industries
Inorganic chemicals and mined substances	Smelting; metallurgical industry; chemical industry; acid mine drainage

III. WATER POLLUTION CONTROL TECHNOLOGIES

Similar to air pollution control, water pollution control is a major discipline with a very large number of professionals engaged in research, development, monitoring, testing, and application of water pollution control techniques. The available literature is obviously very vast and the following discussion is but a brief introduction.

Table 5 lists some major sources and pollutant categories associated with them. The sources of water pollution are numerous. Just to provide one example, a 1000-MW coal-fired power plant can generate up to 200,000 tons of ash each year. This ash contains numerous toxic elements (chromium, cadmium, etc.), although in small concentrations. This ash pile, when rained upon, leaches out the toxic and other chemical components which reach water bodies either through surface runoff or by seepage through the soil to the groundwater which could eventually reach lakes and rivers. The sources of water pollution can be categorized into point sources and nonpoint sources as follows:

Table 6
WATER USE CLASS AND AVAILABLE REGULATION

Use class	Available regulation
Drinking water or community water	Primary standards per EPA 40 CFR 141
Recreation and aesthetics	Secondary standards per EPA 40 CFR 143
Aquatic and wildlife use	EPA water quality criteria and state regulations as appropriate
Agricultural use	No specific federal standard, states set up their own standards
Industrial use	No specific federal standard, states set up their own standards

A. Point sources of water pollution examples

1. The largest volume single point source is the municipal wastewater which consists of domestic sewage and industrial sanitary sewage.
2. The second largest point source of water pollution is industrial wastewater consisting of the products of industrial production contaminated during use as process water and from in-plant spills.

B. Nonpoint sources of water pollution examples

1. Leachates from hazardous waste dump sites.
2. Leachates from clandestine ("midnight") burial sites.
3. Direct discharge of industrial wastes to the waters without authorization or permits ("midnight dumping").
4. Accidental dumping of toxic wastes, including pesticides, directly to the waters.
5. Accident spills from traffic accidents involving transport of hazardous chemicals.

Water pollution control technologies can effectively clean up only the first category of wastes (Category A); for Category B, the technologies are costly and have a lower probability of successful cleanup. The Category B sources should be eliminated by prevention, rather than cured by using control technologies.

A. Water Quality Standards

Cleaning up the wastewaters in Category A can be done only in relation to a set of standards and also in relation to the intended end use of the cleaned up water. It should be remembered that even in unpolluted geographical areas, rain water, when collected and analyzed, is found to contain dissolved CO_2, N_2, and some suspended dust particles. Surface and well waters contain dissolved compounds of metals like Na, Mg, Ca, and Fe. Therefore, water quality should be looked upon as that quality in which no substance is present in the water in concentrations sufficient to prevent its use for the intended normal purposes. There are at least five types of uses of water as indicated in Table 6.

The EPA primary (40 CFR 141) secondary (40 CFR 143) water quality regulations specify maximum contaminant levels (MCL) for most pollutant categories shown earlier in Table 5. The existing EPA regulations cover approximately 35 contaminants which include the traditional biological contaminants, coliforms and turbidity; 13 inorganic contaminants; six organic pesticides; radionuclides; and trihalomethanes. The primary standards are considered to be directly related to public health concerns, the secondary standards result from aesthetic

Table 7
SUMMARY OF EPA PRIMARY AND SECONDARY DRINKING WATER REGULATIONS, MAXIMUM CONTAMINANT LEVELS

	Primary (40 CFR 141)	Secondary (40 CFR 143)
Inorganic chemicals (mg/ℓ)		
Arsenic	0.05	
Barium	1.0	
Cadmium	0.01	
Chromium	0.05	
Copper	None[a]	1.0
Iron	None	0.3
Lead	0.05	
Manganese	None	0.05
Mercury	0.002	
Nitrate (as N)	10	
Selenium	0.01	
Silver	0.05	
Fluoride	4.0	2.0
Zinc	None	5.0
Organic Chemicals (mg/ℓ)		
Endrin	0.0002	
Lindane	0.004	
Methoxychlor	0.1	
Toxaphene	0.005	
2,4-Dichlorophenoxy-acetic acid	0.1	
2,4,5TP (Silvex®)	0.01	
Total trihalomethanes	0.01	
Turbidity (turbidity units, TU)	1 to 5 TU	
Coliform bacteria (counts/100 mℓ)	1 to 4[b]	
Radioactivity (pCi/ℓ)		
Combined Ra-226 and Ra-228	5[c]	
Gross alpha (excluding radon and uranium)	15[c]	
Tritium	20,000[c]	
Strontium-90	8[c]	
Other contaminants		
Chloride (mg/ℓ)	None	250
Color (color units)	None	15
Corrosivity	None	Noncorrosive
Foaming agents (mg/ℓ)	None	0.5
Odor (threshold odor number)	None	3
pH	None	6.8—8.5
Sulfate (mg/ℓ)	None	250
Total dissolved solids (mg/ℓ)	None	500

[a] None means there are no applicable limits set under primary standards.

[b] Please refer to full original regulations (40 CFR 141, 142, 143) for precise information on implementation. This table is only a summary.

[c] If more than one radionuclide is present, they should not be added because the total radioactivity should be such that the annual dose equivalent to any internal organ or total body shall be less than 4 mrem/year.

considerations. These standards are summarized in Table 7. This table is for purposes of information only, and the original regulations should be referred to for precise information for actual regulatory use.

The EPA continually reviews its regulations to evaluate if additional contaminants need to be regulated. The diligent incorporation of water quality engineering practices which employ effectively the available control technologies could prevent further possible regulations by preventing the contaminant problem before it occurs.

B. Selection of Water Pollution Control Technologies

The scope of this book permits only a coarse treatment of this specialized discipline. Also, as stated previously, we will discuss control technologies for large point sources (municipal and industrial wastewater) and hope that nonpoint sources would be eliminated gradually by strict enforcement of the permit process including penalties for midnight dumping. Also, industrial in-plant liquid wastes are expected to be treated with in-plant control technologies to eliminate or reduce emissions in effluents before discharge to municipal systems by the industry.

The following discussion of control techniques applies to both municipal and industrial wastewater treatment. Figure 7 is a presentation of most of the unit processes and operations employed in wastewater treatment. These processes include the following sequential steps:[10]

- Pretreatment
- Primary treatment
- Secondary treatment
- Tertiary (or advanced) treatment
- Disposal of treated water by returning to the environment
- Disposal of sludge after treatment

The first five steps generate sludges that must be treated and disposed of. Usually, effective sludge treatment is obtained by aerobic or anaerobic digestion. Wet combustion is another very efficient method available. Treated sludges can also be returned to the environment as indicated in Figure 7.

The general principle of the control methods is to first remove small particles (grit) by passing the wastewater through screens, then provide adequate time to settle the impurities in lagoons and settling ponds, collect the clear overflow in subsequent ponds, reduce organics and microorganisms using processes such as oxidation in trickle filters, and eventually, chemically treat the water with chlorine to reduce bacterial contamination. Most municipal wastewater containing domestic sewage is treated along these lines.

Devices available for wastewater treatment include the following:

- Lagoons
- Settling ponds
- Screens
- Flotation tanks
- Vacuum filtration (reduces water in sewage sludge) chemical precipitation (alum helps precipitate impurities)
- Activated carbon treatment
- Wet air oxidation (to combust pulpmill waste, sewage, waste stream containing organic matter that can be burned easily)
- Trickle filters (degradation or organic matter with microorganisms)
- Anaerobic waste treatment (decomposes organic matter in the absence of oxygen)

The type of device and the extent of wastewater treatment to be selected for a particular application is a function of the local regulations, federal regulations, the availability of land area for the device (for example, lagoons require large areas), the quality of water (use class) to which the treated wastewater is to be returned if any, the end use of the treated water, the end use for the treated sludge, and cost effectiveness of the treatment process.

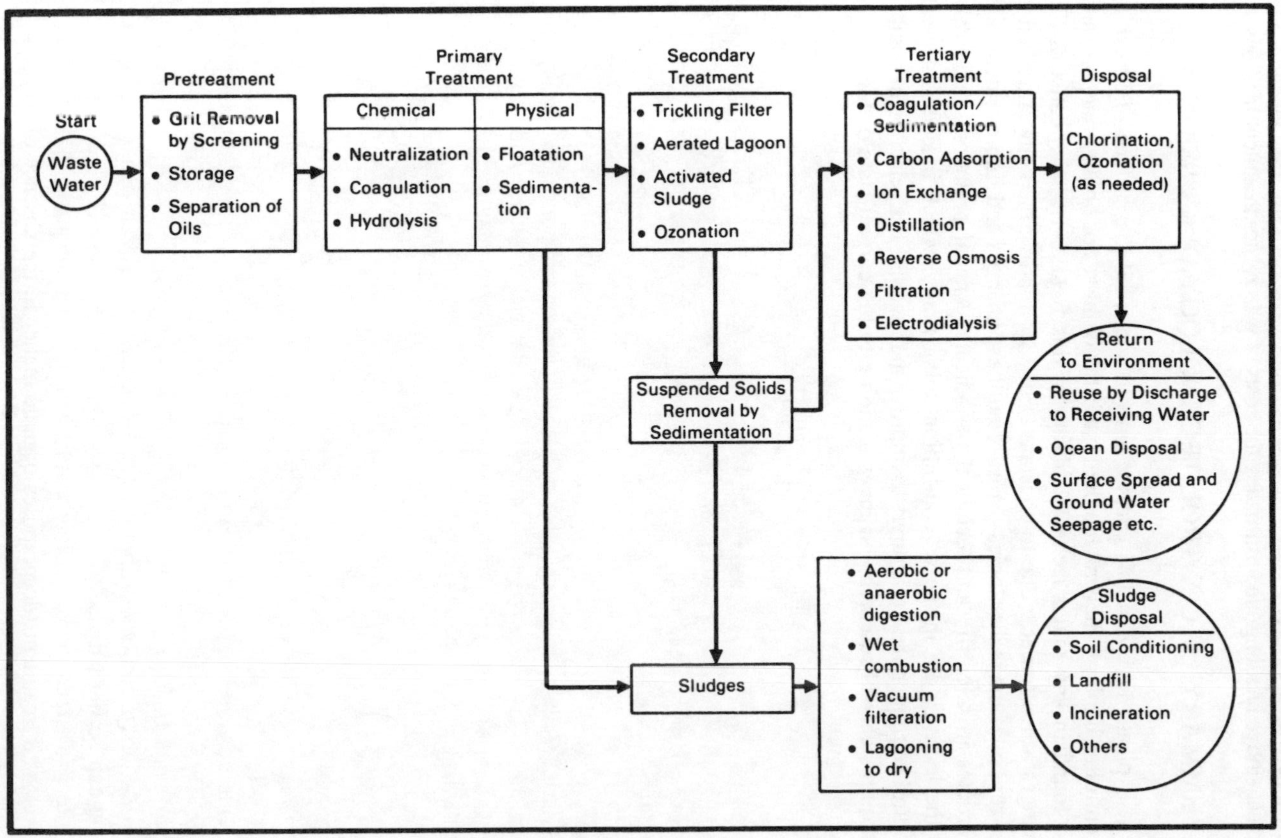

FIGURE 7. Schematic of wastewater treatment unit operations.

Further details of the processes and methods available for treatment of wastewater and sludges can be found in Metcalf and Eddy[20] and Eckenfelder and Santhanam.[10] Benefield et al.[2] describe the chemistry of water and wastewater treatment processes. The *Handbook of Advanced Wastewater Treatment*[8] is an excellent source for understanding the details of the various treatment processes such as coagulation, flocculation, sedimentation, recarbonation, filtration,, activated carbon treatment, disinfection, nitrogen removal, phosphorous removal, chemical sludge handling, and estimating the costs of wastewater treatment facilities.

IV. SUMMARY OF CONTROL TECHNOLOGY DISCUSSION

The reader will please note that this chapter has but scratched the surface of literature available on control technology. Several hundred published volumes of books, manuals, and reports can be found in any major technical library in this field. Also, the status of the technology changes at least yearly due to constant efforts to improve the competitive edge and cost effectiveness of the technologies. Most often these changes and improvements are published as articles in scientific and technical periodicals. Annual pollution control expenditures by the industry are in the several-billion-dollar range, which indeed is a great incentive to control equipment manufacturers to improve their wares. The selection of control technologies requires careful research and investigation into the current state of the art.

Chapter 7

NEPA AND THE NUCLEAR WASTE POLICY ACT

I. INTRODUCTION

In any discussion of the NEPA process, the chief aim of which being to preserve, protect, and improve the quality of the human environment, the subject of nuclear waste disposal cannot be ignored because it is an issue of major concern to the public. Nuclear waste includes wastes generated by commercial nuclear power-generating stations and from national defense nuclear material production activities. In the U.S. today, there are no facilities for the permanent disposal of the high-level radioactive waste generated during the production of electricity in nuclear power plants or during the production of nuclear materials for national defense. Although nuclear activities produce smaller volumes of wastes compared to other major activities that generate hazardous wastes, nuclear wastes are radioactive and therefore require special handling and storage. Nuclear wastes have been safely stored in a temporary manner for decades with no discernible adverse effects on the health and safety of the public. However, some part of them will remain radioactive for thousands of years, and hence the wastes need to be safely disposed.

To address the need for a safe and environmentally acceptable method of permanent disposal of radioactive wastes, the U.S. Congress passed the Nuclear Waste Policy Act of 1982 (NWPA) on December 20, 1982. The act was signed into law by the president on January 7, 1983. This law has three titles:

Title I — Disposal and storage of high-level radioactive waste, spent nuclear fuel, and low-level radioactive waste

Title II — Research, development, and demonstration regarding disposal of high-level radioactive waste and spent nuclear fuel

Title III — Other provisions relating to radioactive waste including establishing a nuclear waste fund

In Section 114(a)D of the NWPA, Congress employed a unique method of exempting parts of the environmental impact statement (for site approval and construction authorization) from selected provisions of the CEQ regulations, as follows:

1. The EIS shall not be required to consider the **need** for a repository.
2. The EIS shall not be required to consider **alternatives** to geologic disposal.

The NWPA itself was considered sufficient to justify adequately the need for the repository. Also, the act specified geologic disposal as the method of high-level waste and spent nuclear fuel disposal.

This Congressional action, to exempt the geological disposal facility EIS from certain provisions of NEPA, is justifiable on the basis that the geologic disposal method was selected after about 10 years of intensive and detailed technical and environmental evaluations of the following different alternatives:[34]

1. Very deep hole waste disposal
2. Rock melt waste disposal
3. Island-based geologic disposal
4. Subseabed disposal

5. Ice sheet disposal
6. Well injection disposal
7. Transmutation
8. Space disposal
9. No action

II. EVOLUTION OF THE U.S. NUCLEAR WASTE DISPOSAL PROGRAM

The exemptions from parts of NEPA requirements can be further explained by the fact the Nuclear Waste Policy Act resulted from planning that began in 1957. The planning activities began with the National Academy of Sciences panel recommendations detailed below.[26]

- 1957 — National Academy of Sciences panel recommends salt deposits be examined as possible repository sites.
- 1960 — Study of disposal in South Carolina bedrock begins.
- 1962 — First industrial processing of high-level waste begins at West Valley plant in New York.
- 1963 to 1967 — Project Salt Vault, Lyons, Kansas.
- 1968 — Evaluation of Hanford basalts begins in the state of Washington.
- 1970 — Atomic Energy Commission issues regulation requiring on-site solidification of liquid high-level radioactive wastes at commercial reprocessing plants within 5 years, and shipment of solidified waste to a federal repository within 10 years.
  — Pilot Facility at Lyons, Kansas
- 1972 — Study of disposal in South Carolina bedrock terminated, AEC abandons Lyons project, proposes alternative waste solution using retrievable surface storage facilities (RSSF), search continues in Kansas, expanded to New Mexico, USGS begins reconnaissance in Paradox Basin in Utah and Colorado.
- 1973 — Subseabed Disposal Program begins.
- 1974 — Waste Isolation Pilot Plant (WIPP) facility proposed for site in New Mexico evaluation of salt bed.
- 1975 — Energy Research and Development Administration, AEC successor, abandons RSSF concept in favor of broader approach to waste management.
- 1976 — National Waste Terminal Storage (NWTS) Program announced; Office of Waste Isolation (OWI) formed at Union Carbide; evaluation of geologic formations at Nevada Test Site (NTS) begins; evaluation begins of Permian basin in Texas, excluding Delaware subbasin; President Ford defers commercial reprocessing of nuclear spent fuel during election campaign against Carter, but calls for demonstration of commercial high-level waste repository by 1985.
- 1977 — Carter Administration indefinitely defers commercial reprocessing; Department of Energy, ERDA successor, announces plan for U.S. to take spent fuel from utilities and store it at away-from-reactor sites until permanent disposal method is selected; evaluation begins of Salina salt basin in Michigan, New York, and Ohio; evaluation of Palo Duro subbasin (Permian basin) begins.
- 1978 — Study areas recommended in Gulf Coast and Salina salt regions; Office of Nuclear Waste Isolation (ONWI) formed at Battelle Memorial

Institute, Columbus, Ohio; Interagency Review Group on Nuclear Waste Management recommends establishment of several geologic repositories for spent fuel and high level waste, but abandons 1985 target date. Instead, early small-scale demonstration of geologic burial of spent fuel rods is proposed, possibly at Waste Isolation Pilot Plant (WIPP) in New Mexico.

- 1979 — Draft environmental impact statement (U.S. DOE 1980)[34] recommends geologic repositories as solution for high-level waste disposal from among ten alternative approaches such as burial in polar ice caps or ocean beds. Congress appropriates funds for WIPP plant in New Mexico, but only for nuclear waste not subject to NRC licensing requirements.
- 1980 — President Carter announces U.S. will evaluate four to five sites in different geologic media before selecting site for first permanent repository. Carter opposes early demonstration at unlicensed WIPP facility and establishes State Planning Council to recommend procedures for selecting repository sites. Date for operation of first commercial geologic repository slips to 1999 to 2006 period. Congress fails to pass Nuclear Waste Policy Act that would expedite Carter Administration program, over issue of whether to license defense nuclear wastes. Congress enacts legislation permitting states to enter into regional compacts to choose sites for burial of low-level waste.
- 1981 — The U.S. Department of Energy (DOE) announces it will proceed with WIPP facility as repository for transuranic waste and as unlicensed experimental facility for small quantities of defense high-level waste.
  — Reagan Administration lifts indefinite ban on commercial reprocessing of spent fuel. Department of Energy expedites Carter Administration program and plans for operation of Test and Evaluation Facility by 1988 and first full-scale geologic repository by the mid-1990s.
- 1982 — Congress passes the Nuclear Waste Policy Act of 1982.

The key feature of the Nuclear Waste Policy Act of 1982 is the provision for the development of deep geologic repositories for the disposal of high-level radioactive waste and spent nuclear fuel. The act specifies a schedule and a step-by-step process by which the president, the Congress, the states, affected Indian tribes, local governments, and federal agencies must collaborate in the siting, design, construction, and operation of the geologic repositories. The act has provided a mandate and a set of rules — including unprecedented interaction among the federal government, the states, and the public, as well as compliance with NEPA requirements — for proceeding with this program.

III. THE PROCESSES OF NUCLEAR WASTE GENERATION

In 1984, 86 domestic nuclear power plants generated more than 13% of the electricity produced in the U.S., making nuclear power the second leading energy source of electricity generation in the U.S., following coal. Another 30 nuclear plants are either under construction or planned. Recent government studies predict that nuclear power will play an expanding role in the supply of electricity in this country and will generate about 20% in the 1990s. However, the growth of the nuclear power industry beyond the 1990s is uncertain for a variety of economic and institutional reasons.[6]

For each large (1000-MW electric) nuclear power plant, about 33 tons of uranium oxide fuel, which has been gradually irradiated during the fission process, are removed from the power plant reactor each year. At this point, the fuel can no longer sustain an efficient chain reaction and is extremely radioactive and thermally hot. It is referred to as "spent fuel". The fuel consists typically of pellets of ceramic uranium oxide sealed in hundreds of metal rods which are bundled together within a very strong, rigid metal structure called a "fuel assembly". The fuel rods are carefully spaced in the fuel assemblies to allow coolant (water in most cases) to flow between them. Each assembly is about 12 to 14 ft long and weighs about 1200 lb. Fuel assemblies are designed to be readily handled with suitable hoists and cranes at the reactor sites. Several hundred fuel assemblies are closely packed vertically into what is called the "reactor core".

Spent fuel assemblies that are no longer suitable for operations in the reactor are removed and stored in a deep water pool adjacent to the reactor. The water cools the spent fuel, which is thermally hot, and serves as an effective shield to protect workers at the reactor site from high-energy gamma radiation. The water is kept free from minerals that would corrode the fuel rods. It remains in that storage pool and is not discharged, although it is replenished and filtered.

The level of radiation within the fuel assemblies begins to decline immediately after their removal from the reactor core, and within 10 years, it has decayed by 90%. Nevertheless, technicians constantly inspect the spent fuel assemblies to ensure safety. Storage of spent fuel in water basins is a temporary measure; the spent fuel will ultimately be shipped to geologic repositories for permanent disposal.

Beginning about 1956 through the end of 1984, spent fuel assemblies containing about 10,000 metric tons of uranium (MTU) have been discharged (taken out of service) from commercial nuclear power plants. This quantity of discharged fuel would be approximately equal to the volume of one football field 3 ft deep — a relatively small size. The inventory of spent fuel will grow and may approach 130,000 metric tons by the year 2020. This amount is still very small in physical quantity in comparison with the wastes from coal-burning electricity generation plants. A single-modern coal-fired electricity plant, for example, produces several thousands of tons of ash, gases, and sludge each year that must be properly controlled and disposed of in accordance with strict environmental regulations.

A small quantity of liquid high-level radioactive waste was generated during the commerical reprocessing of reactor spent fuel at a facility near West Valley, New York, from 1966 through 1972. (Reprocessing is a chemical separation process that can extract plutonium, which is created during the fission process, and the remaining usable uranium from spent fuel rods. These energy-containing elements can be fashioned into new fuel assemblies for reinsertion into reactors to produce more energy.) This liquid waste is scheduled to be solidified into glass and encapsulated in stainless steel canisters for eventual disposal in a geologic repository, similar to the spent fuel.

Another significant source of radioactive waste needing permanent disposal is the defense high-level radioactive liquid waste (DHLW) produced by the national atomic energy defense production activities. The DHLW includes the highly radioactive liquid waste material that results from the reprocessing of specially designed and irradiated nuclear fuel to obtain nuclear materials for weapons. Defense high-level waste is generated and stored at three U.S. Department of Energy (DOE) sites: the Savannah River Plant (South Carolina), the Idaho National Engineering Laboratory (Idaho), and the Hanford site (Washington). These wastes will be processed to convert them to a solid form (immobilized in a solid matrix), stored on site temporarily, and shipped to the geologic repository (when it is constructed under the NWPA) for permanent disposal.

The spent fuel and the liquid radioactive waste from reprocessing contain fission products generated in nuclear reactors that remain radioactive and hence require isolation for up to 10,000 years. For this reason, the federal government seeks to implement a permanent, viable solution to the waste disposal problem.

IV. A NATIONAL POLICY FOR MANAGING NUCLEAR WASTE

After numerous scientific and technical studies to identify a safe disposal method, a national policy for the management of nuclear wastes emerged over the decades. National programs for nuclear waste management that had operated under very general guidelines from Congress became focused in the 1970s and early 1980s. President Reagan's "Nuclear Policy Statement" of October 8, 1981 contributed significantly to the forging of a national consensus on a safe radioactive waste management policy. This national consensus led to the passage by Congress, on December 20, 1982, of the first comprehensive legislation for the management of high-level radioactive waste in the history of the U.S.

Successful implementation of the NWPA requires the resolution of substantial technical and institutional challenges. First, the public must be convinced that the interests of the nation are best served by consolidating radioactive waste now stored temporarily at nearly 100 sites around the country in one or two geologic repositories designed for permanent disposal. Next, the DOE must demonstrate to the citizens of the country that the waste disposal program will be safe, efficient, and timely. Furthermore, the DOE must balance conflicting and competing interest, as reflected by the tension between a rigorous, mandated program schedule and the need for extensive public and community involvement.

The act requires the DOE to obtain a license from the U.S. Nuclear Regulatory Commission (NRC) to construct and operate geologic repositories for spent fuel and high-level waste in a manner that "will provide a reasonable assurance that the public and the environment will be adequately protected . . . ." In addition to establishing a detailed schedule for the siting and licensing of one or more geologic repositories, the act contains several other features, including the following:

1. Assigning responsibility for the full payment of disposal costs to those who benefit from those services (i.e., the owners and generators of high-level waste and spent fuel) and, accordingly, creating in the U.S. Treasury a special nuclear waste fund to be composed of the payments made by those using the services. These user fees are paid by rate payers who use nuclear-generated electricity.
2. Committing the federal government to study monitored retrievable storage in parallel with permanent geologic repositories.
3. Providing for a federally owned and operated system of interim storage facilities for no more than 1900 metric tons of spent fuel from civilian reactors where the owners of which cannot reasonably provide adequate storage capacity at the reactor site and creating an interim storage fund to make certain that those using the services will pay the full cost.
4. Setting out an extensive system of checks and balances through public involvement, along with written consultation-and-cooperation (C&C) agreements, to ensure federal accountability to and communications with states, affected Indian tribes, and the public.

V. THE GEOLOGIC REPOSITORY

The geologic repository will combine two types of industrial facilities: (1) a nuclear waste processing and handling facility at the surface and (2) a large mine constructed 1000 to 4000 ft below the surface. For each repository, the underground area which is designed primarily for the emplacement of 70,000 metric tons (uranium) of spent fuel, will cover approximately 2000 acres. Above the ground, a central area of about 400 acres will contain a variety of buildings and structures that will be used during the operations period of the repository, which should be about 30 years. Rail and truck unloading areas, warehouses and administration buildings, water treatment plants, a storage area of excavated rock, and other facilities will comprise the central complex, in addition to the waste handling building. A restricted surface zone slightly larger than the area of the underground repository will be established. After operations have ceased, each of the two repositories will be sealed and decommissioned, with all reasonable efforts taken to alert future generations of its existence, including the installation of highly visible markers denoting such prohibitions. The Department of Energy is investigating a range of possible measures, in addition to the markers, that could be used to inform future generations about the repository while the emplaced radioactive waste decays to safe levels.

VI. SITE SELECTION

The process of selecting a suitable site for each repository is dictated by the requirements of the NWPA and influenced by the extensive research and development activities that preceded enactment of that legislation. The DOE search for an acceptable first repository site include several factors. Among these are the types of geologic formations to be considered, the geologic history of those formations, the public safety and socioeconomic concerns, and the environmental consequences. Based on lengthy analyses of these factors, DOE, in February 1983, identified nine potentially acceptable sites in six states for the first repository. These sites are located in the following geologic formations:

- Salt — natural sodium chloride in rock form
- Basalt — solidified volcanic lava
- Tuff — solidified volcanic ash deposits

These initial selections were fundamental to the DOE decision in December 1984 to propose three of these sites as the preferred ones for recommendation to the president for site characterization. These sites are

- Nevada — Yucca Mountain site — tuff geology
- Texas — Deaf Smith County site — salt geology
- Washington, DOE Hanford Reservation site — basalt geology

The act requires the DOE to begin operating the first repository by January 31, 1998. Since there have already been scheduled delays, the DOE may be ready to accept the wastes by January 31, 1998, even if the repository will not be fully operational. The second repository may not be built at all according to present developments. The host state or any affected Indian tribe may issue a notice of disapproval regarding construction authorization for a repository. The act allows for a Congressional override in the event a site previously recommended by the president is vetoed by any affected party. The DOE is striving to develop a credible, technically sound program that will gain the support of all parties involved.

Although the act stipulates the schedule and conditions of the site selection process, public perception of the benefits and weaknesses of a geologic repository and the resulting public opinion could influence the final decision on construction and operation of the first geologic repository.

Section 8 of the NWPA requires the president to evaluate the use of the geologic repository for the disposal of defense high-level radioactive waste disposal. After conducting such an evaluation, using factors relating the cost, efficiency, health and safety, regulations, transportation, public acceptability, and national security, the president decided on April 30, 1985, that defense-generated wastes do not require defense-only repositories and defense wastes can be disposed of in geologic repositories built for commercial wastes.

The NWPA has three other key provisions: (1) the monitored retrievable storage program (MRS), (2) the interim storage program, and (3) nuclear waste fund.

VII. THE MRS FACILITY

The purpose of a MRS facility, which would be licensed by the NRC, is to receive, consolidate, package, and subsequently store spent fuel assemblies in either sealed storage casks or dry wells before they are shipped to geologic repositories for disposal. The packaged spent fuel assemblies would be sent directly to a repository by rail. The MRS facility would serve as a logistics hub for the transportation system, increase the flexibility of the waste management system by providing a buffer between the acceptance of nuclear waste and its emplacement in the geologic repository, simplify the design, construction, and operation of repository facilities by controlling the flow of spent fuel to the repository, and provide a focal point for system integration activities.

Section 141 of the NWPA directs the DOE to complete a detailed study of the need for the MRS facility and to submit to Congress a proposal for its construction. The Congress exempted the need to prepare an EIS for the proposal, but directed that an environmental assessment shall be prepared and submitted with the proposed.

If the Congress by law, after review of the proposal submitted by the DOE, specifically authorizes construction of a monitored retrievable storage facility, the requirements of the National Environmental Policy Act of 1969 (42 U.S.C. 4321 et seq.) shall apply with respect to construction of such facility, except that any environmental impact statement prepared with respect to such facility shall not be required to consider the need for such a facility or any alternative to the design criteria for such facility. Here, again, the Congress provided specific exemptions from NEPA requirements. There is not violation of the spirit or letter of NEPA in these exemptions because the need and alternatives were carefully analyzed before the proposal for MRS was submitted to Congress.

VIII. THE INTERIM STORAGE PROGRAM

The purpose of an interim storage program, authorized under Section 131 of the NWPA, is to relieve the shortage of spent fuel storage space at commercial nuclear power plants by providing interim storage for not more than 1900 metric tons of spent fuel. Environmental documentation requirements for this program are based on whether the storage facility (1) would hold up to 300 metric tons of spent fuel (in which case the NWPA does not classify it as a major federal action) or (2) hold more than 300 metric tons (in which case it is a major federal action). Accordingly, the DOE shall prepare and make available to the public an environmental assessment of the probable impacts of any provision of less than 300 metric tons of storage capacity at any one federal site under subsection (a)(1)(A) that requires the modification or expansion of any facility at the site, and a discussion of alternative activities that may be undertaken to avoid such impacts. Such environmental assessment shall include

1. An estimate of the amount of storage capacity to be made available at such site.
2. An evaluation as to whether the facilities to be used at such site are suitable for the provision of such storage capacity.
3. A description of activities planned by the secretary with respect to the modification or expansion of the facilities to be used at such site.
4. An evaluation of the effects of the provision of such storage capacity at such site on the public health and safety and the environment.
5. A reasonable comparative evaluation of current information with respect to such site

and facilities and other sites and facilities available for the provision of such storage capacity.

6. A description of any other sites and facilities that have been considered by the secretary for the provision of such storage capacity.
7. An assessment of the regional and local impacts of providing such storage capacity at such site, including the impacts on transportation.

The provision of 300 or more metric tons of storage capacity at any one federal site will require the preparation of an EIS, as defined in the CEQ regulations (40 CFR 1500). Both the EA and EIS are subject to judicial review.

IX. THE NUCLEAR WASTE FUND

Implementing the NWPA provisions could cost over $20 billion (1984). How will this cost be paid? There again the Congress was creative. Section 302 of the NWPA stipulates that the owners and generators of the nuclear waste who enter into service contracts with the federal government will pay the full costs of waste disposal. To cover the costs for disposing of commercial spent nuclear fuel, utilities with nuclear power plants are obliged to pay two fees — one for electricity produced by nuclear power on or after April 7, 1983, and another one for spent fuel discharged prior to April 7, 1983. The former fee was set initially by the Congress at 1.0 mil/kWh, while the latter one-time fee was in the form of dollar per kilogram charges that equate to 1.0 mil/kWh.

The major source of funds for the waste disposal program is the ongoing fee, which is a constituent part of the nuclear waste fund. Revenues from the collection of the ongoing fee are rising steadily. Current revenues exceed $300 million a year; and they are expected to reach $500 and $600 million annually within a few years. The one-time fee comprises the other component of the nuclear waste fund. DOE estimates the one-time fee will eventually produce revenues of over $2.3 billion.

Provisions of the act require the secretary of energy to review annually the user fees to determine if they cover program costs. The DOE has published three nuclear waste fund (NWF) fee assessment reports since 1983. The report published in February 1985 concluded that the current fee of 1.0 mil/kWh is adequate at this time for a reasonable range of program costs, nuclear generation, and economic forecasts.

Enactment of the Nuclear Waste Policy Act of 1982 represented the culmination of an extended national effort to resolve the problems associated with the management of nuclear waste. The responsibilities vested in the U.S. Department of Energy to develop a cogent, technically sound and environmentally acceptable waste management program are significant. The exemptions that the Congress wrote into the NWPA from certain requirements of NEPA in the preparations of an EA and an EIS are significant, but justified on the basis of the large efforts expended to understand the environmental consequences before the NWPA legislation was written.

Chapter 8

AFTER THE RECORD OF DECISION

I. INTRODUCTION

As one approaches the tail end of the NEPA process compliance effort on a project, it is not unusual to feel that once an ROD is published, the task is done and one can rest, particularly after the long and tedious effort that was needed to conduct the environmental analysis and to publish the DEIS, FEIS, and the ROD. This feeling that all is well after the ROD must be tempered by the understanding that for controversial projects, the ROD is not the finale of the NEPA process in the public empathy aspects. Even after the ROD is published, the lead agency can face

- Organized demonstrations by environmental groups to stop the project.
- Court actions by concerned environmental groups to stop the project.
- Inquiries by news media, particularly investigative reporters, on how the EIS process was conducted and if the commitments made in the EIS are being satisfied.
- Press interviews suggesting the NEPA process was not professionally conducted.

The press interviews can occur several years after the ROD publication, when in the normal course of business, the EIS team will have moved away from the EIS project, and taken on other responsibilities. While the press reporter can easily ask questions, answering the questions, particularly on specific technical issues, takes a bit of preparation and revisiting the EIS. Therefore, the lead agency is well advised to make sure that the questions from the press are provided well in advance to the EIS project manager to allow time to study the questions and prepare adequate answers. At least three working days should be allowed. The lead agency public relations staff should brief the EIS team on whether the press interview is anticipated to be benign or menacing, general or specific in terms of technical content, and brief or protracted. This type of data will prepare the EIS team psychologically for the interview.

II. RECORD KEEPING

For reasons stated above, each lead agency needs to develop a method of preserving the records of the EIS project for at least 5 years after the publication of the ROD. On very large and controversial projects in which the implementation period can exceed 10 years, the records may have to be preserved for 10 years or longer. The original EIS staff may have moved on to other jobs, left the organization, or be unavailable for other reasons. Therefore, well preserved and understandable records are the only method of preserving the memory of EIS projects and of the process involved in preparing the EIS and ROD.

To ensure that useful records are prepared during the course of the EIS project, each large EIS project may create a position of an EIS records keeper. The position description would be to ensure that a registry of all auditable records, data sources, and correspondence is maintained, filed, and stored for later retrieval. Such a clearly defined separate position is necessary because very often, EIS project managers and staff are so involved in day-to-day technical details and delivery deadlines, that they fail to keep accurate record of all the data. For up to 10 years after an EIS is published and ROD is implemented, the fourth estate can hound the EIS team. The best way to be prepared is to keep good records.

Examples of types of records to be saved for retrieval are

1. Notebooks in which all assumptions and methods of calculating environmental impacts are written out and defined.
2. Correspondence with organizations who provided raw data for impact calculations and for description of alternatives.
3. Results of EIS team review of raw data for quality and consistency.
4. Results of audits for quality assurance and quality control, if any.
5. All correspondence between lead agency and EIS contractor.
6. All correspondence between EIS contractor and applicant for license/permit, etc.
7. All correspondence with any party who provided technical or legal support for the EIS.
8. A record of telephone conversations that led to decisions and data transmission, memos of meetings, and other notes that contain data used in preparing the EIS.

III. MITIGATION COMMITMENTS IN RECORD OF DECISION

Depending on the level of the NEPA documentation adopted for a project (EA, FONSI or EIS, ROD), enforceable mitigation measures, monitoring programs, or other requirements can be specified in FONSI or ROD. Such enforceable requirements need to be followed through very carefully to ensure that they are indeed incorporated in the implementation of the proposed action.

By definition, mitigation includes the following according to 40 CFR 1508.20:

(a) Avoiding the impact altogether by not taking a certain action or parts of an action.
(b) Minimizing impacts by limiting the degree or magnitude of the action and its implementation.
(c) Rectifying the impact by repairing, rehabilitating, or restoring the affected environment.
(d) Reducing or eliminating the impact over time by preservation and maintenance operations during the life of the action.
(e) Compensating for the impact by replacing or providing substitute resources or environments.

Additional examples of mitigation can be added, as follows:

(f) Reducing or eliminating the impacts by use of best available control technology with due regard to cost benefits.
(g) Monitoring the emissions so that careful watch can be placed to identify the impact level at which plants may require temporary shut down.

IV. AUDIT OF COMMITMENT COMPLIANCE

When commitments are made in the EA, FONSI, EIS, or ROD to adopt a mitigation measure, such commitments should be defined in very clear and specific terms in the NEPA document so that there is no ambiguity as to the details of the mitigation measure. About 1 year after the ROD is published (or FONSI is published as appropriate), the EIS team and particularly the lead agency EIS team, should record on how the mitigation measure was implemented in the proposed action. Another important measure to ensure that mitigation commitments are implemented is to audit the effectiveness of the measure during the operation of the facility. The procedures for the audit process should be defined by the lead agency and its compliance should be periodically verified by the lead agency.

Chapter 9

EMPATHIZING WITH THE PUBLIC

I. INTRODUCTION

The importance of the EIS team to identify with and understand the concerns, feelings, and motives of the public who provide comments during the scoping process and on the draft EIS cannot be overstated. Any comments received during the entire NEPA process should be carefully considered in preparing the NEPA documents.

It was not too long ago that smoke stacks were welcomed as a sign of industrial prosperity around our towns and brought gladness to the hearts of the residents. Beginning in the 1960s, this attitude changed drastically because the public began to realize that smoke is not so innocent as it looks, and it pollutes the air slowly and sometimes irreversibly. The often unseen waste discharges to the water resources of the community kill fish, clog the waters' capacity to rejuvenate itself, and make the waters smelly and unfit for recreational use.

Evidence of these insulting effects of poorly planned industrialization was becoming hard to ignore for the public. Several industrial facilities continued to ignore the evidence and thereby created the perception of belittling the public concern regarding the pollution of the environment. The quality of the air was definitely deteriorating. There was much public outcry via demonstrations and other forms of expression. The Congress took notice and the National Environmental Policy Act was enacted in 1969.

The Congress recognized that the availability of reliable technical information to the public and governmental regulatory bodies on the nature of environmental insults from proposed new industrial facilities was the key to making informed decisions to protect the environment while not throttling needed industrial growth. Accordingly, Section 102(2)(G) of the act stipulated that "all agencies of the Federal government shall make available to states, counties, municipalities, institutions, and individuals, advice and information useful in restoring, maintaining, and enhancing the quality of the environment." This provision of the act forces the federal agencies to provide information to the public on all activities (including licensing and permitting of new industrial facilities) that have the potential to impact effects on the environment, safety, and health of the public.

The clearest directive on public involvement to agencies is contained in 40 CFR 1506.6 issued by the CEQ and is summarized below.

Agencies shall

(a) Make diligent efforts to involve the public in preparing and implementing their NEPA procedures.

(b) Provide public notice of NEPA-related hearings, public meetings, and the availability of environmental documents so as to inform those persons and agencies who may be interested or affected.

(c) In all cases the agency shall mail notice to those who have requested it on an individual action.

(d) In the case of an action with effects of national concern, notice shall include publication in the *Federal Register* and notice by mail to national organizations reasonably expected to be interested in the matter. An agency engaged in rule making may provide notice by mail to national organizations who have requested that notice regularly be provided. Agencies shall maintain a list of such organizations.

(e) In the case of an action with effects primarily of local concern, the notice may include:

(i) Notice to state- and area-wide clearing houses.
(ii) Notice to Indian tribes when effects may occur on reservations.
(iii) Following the public notice procedures of the affected state for comparable actions.
(iv) Publication in local newspapers (in papers of general circulation rather than legal papers).
(v) Notice through other local media.
(vi) Notice to potentially interested community organizations including small business associations.
(vii) Publication in newsletters that may be expected to reach potentially interested persons.
(viii) Direct mailing to owners and occupants of nearby or affected property.
(ix) Posting of notice on and off site in the area where the action is to be located.

(f) Hold public hearings or public meetings whenever appropriate.
(g) Solicit appropriate information from the public.
(h) Explain in its procedures where interested persons can get information or status reports on environmental impact statements and other elements of the NEPA process.
(i) Make environmental impact statements, the comments received, and any underlying documents available to the public pursuant to the provisions of the Freedom of Information Act (5 U.S.C. 552), without regard to the exclusion for interagency memoranda where such memoranda transmit comments of federal agencies on the environmental impact of the proposed action.
(j) Materials to be made available to the public shall be provided to the public without charge to the extent practicable, or at a fee which is not more than the actual costs of reproducing copies required to be sent to other federal agencies, including the Council of Environment Quality.

II. WHO IS THE PUBLIC?

The public that comments on NEPA documents are of four kinds: (1) any citizen who feels directly or indirectly affected by the action, (2) environmental groups and organizations, (3) trade associations, and (4) federal and state or local governmental agencies, other than the lead agency. The following are some examples:

1. If a large, new shopping center is proposed next to your residential neighborhood, such that new roads increase the traffic all around your home, naturally you would be inclined to oppose the action as an individual citizen. The concern is real and the licensing agency for the action will be well advised to respond to your concern.
2. Environmental organizations are by far the most active private groups that provide serious and detailed comments on NEPA documents. These groups are often called ''intervenors'' because they intervene to halt actions through organized protests, court actions, and other effective methods. There are several major environmental groups in the U.S., as listed below:

   - National Wildlife Federation
   - National Audubon Society
   - Sierra Club
   - Wilderness Society
   - Izaak Walton League
   - Environmental Defense Fund
   - Natural Resources Defense Council (NRDC)

- Defenders of Wildlife
- Friends of the Earth

The Sierra Club was instrumental in instituting the PSD permit process by challenging the EPA in a litigation which reached the Supreme Court in which the Sierra Club won its case. Eventually the Congress made the PSD requirement a part of the Clean Air Act. This action by the Sierra Club is an excellent example of the good work that private environmental organizations do in protecting the national environmental quality.

3. Trade associations are usually not active in the environmental protection area. But, when a particular action has the potential to threaten the marketability of a product (for example, the location of nuclear waste repository in a farming area could be perceived by the farming community as a source of radioactivity that would contaminate crops), the appropriate trade associate would intervene to prevent the action or alleviate the perceived potential impacts on their products.
4. The fourth group are governmental agencies who are required to comment on environmental documents. These include the U.S. EPA, state environmental protection agencies, Fish and Wildlife Service, National Park Service, Bureau of Reclamation, and others. Lead agencies are not considered the public, obviously.

III. HOW TO EMPATHIZE

Responsiveness to public concerns can be demonstrated in several ways. The primary requirement in empathizing with the public is for the EIS team to feel and understand the concerns of the public in a sincere spirit. If this occurs, then a well-designed response plan can be effectively implemented. Such a plan may have the following elements:

1. Launch a public information campaign, such as open houses, workshops, and public hearings.
2. Prepare a public comment response plan.
3. Include responses to the public comments in the draft and final EIS.

In launching a public information campaign, it is necessary to not only send press announcements to newspapers and television stations on impending hearings and on availability of documents for public review, but also hold informational workshops to describe the technical aspects of the project and the NEPA document to the public. Also, holding public hearings helps those segments of the public who are not willing to write down their comments to offer oral comments which can be recorded by a qualified stenotypist, like a court reporter.

The preparation of a public response plan was discusssed in earlier chapters. Logging the comments and including such a record of the comments in the EIS shows to the public that each comment was considered. Also, often it is very helpful to include the reproductions of original comment letters in the final EIS so that commenters can see their own letters in the public documents. This is a good way to show that earnest efforts were exerted by the lead agency in being fully responsive to the public opinion.

IV. INFORMATIONAL WORKSHOPS

On major federal or state proposals of intense public concern, workshops during the very early stages of the EIS preparation and on the draft EIS can be very helpful in preventing major public opposition to the project. These workshops can be designed to provide the public an opportunity to ask questions either on the proposed action or on the draft EIS or on the scope of the draft EIS. Workshops help the public to discuss their concerns directly with the lead agency technical staff and the staff of EIS contractor, if any. Workshops can have any workable agenda and format, but usually start off with a short, narrated slide show, an introduction of all lead agency managers or well-known local officials, and include detailed, small group discussions with the public. Such workshops can assist the public in preparation for participation in public hearings.

Workshops help to

- Provide the public an opportunity to become involved in the decision process of the agency.
- Build public trust and confidence in the decision making process of the agency.
- Assist the public to provide informed, meaningful input to the draft environmental impact statement, final environmental impact statement, etc.
- Identify public concerns in order for the lead agency to effectively prepare for public hearings.
- Provide the EIS staff experience and confidence in interacting with the public.

One method of insuring that effective workshops are conducted is to institute formal evaluation of each workshop session to provide feedback to the workshop staff so that future workshops are conducted with improved expertise and confidence. Evaluation can be conducted by

1. Using questionnaires for workshop participants.
2. Telephone interviews with workshop participants.
3. Monitoring media coverage before and after workshop to evaluate any change in reporting, particularly with reference to technical accuracy, bias, and attitude towards the proposed project.

Each workshop staff team can design workshop questionnaires so that they receive the response needed for evaluating the effectiveness of their efforts to reach a large audience, to inform the audience, and gain their confidence in the decision process by showing sincere interest in getting input to the decision process from the public. Reaching a large audience in a passive manner is possible by using exhibits at shopping malls. Such exhibits can draw real attention to impending active workshops and public hearings.

The decision making process is changing into a real democratic way not only because of NEPA, but also as a reflection of the times. Regulatory agencies, like the NRC, the U.S. EPA, FERC, etc., are taking NEPA compliance very seriously. Agency decision makers and executives in industry can eliminate costly project delays by following the procedures outlined here and perhaps design procedures better suited to their specific project needs. In any case, empathizing with the public is as important as the scheduling, costing, and the engineering design of projects.

Chapter 10

ANSWERS TO SOME COMMON QUESTIONS ABOUT NEPA PROCESS

I. INTRODUCTION

The Council on Environmental Quality has published answers to "forty most asked questions concerning CEQ's National Environmental Policy Act Regulations" in *Federal Register,* 46, 55, 1981. So long as NEPA enforcement is active, the answers to these questions will be important information for professionals involved in NEPA documentation. Therefore, these questions and answers are provided below for the information of the reader. The subjects covered by the 40 questions are as follows:

1. Range of alternatives
2. Alternatives outside the capability of applicant or jurisdiction of agency
3. No action alternative
4. Preferred alternative of the agency
5. Proposed action vs. preferred alternative
6. Environmentally preferable alternative
7. Difference between sections of EIS on alternatives and environmental consequences
8. Early application of NEPA
9. Applicant who needs other permits
10. Limitations on action during 30-day review period for final EIS
11. Limitations on actions by an applicant during EIS process
12. Effective date and enforceability of the regulations
13. Use of scoping before notice of intent to prepare EIS
14. Rights and responsibilities of lead and cooperating agencies
15. Commenting responsibilities of EPA
16. Third party contracts
17. Disclosure statement to avoid conflict of interest
18. Uncertainties about indirect effects of a proposal
19. Mitigation measures
20. Worst case analysis
21. Combining environmental and planning documents
22. State and federal agencies as joint lead agencies
23. Conflicts of federal proposal with land use plans, on policies and controls
24. Environmental impact statements on policies, plans or programs
25. Appendices and incorporation by reference
26. Index and keyword index in EISs
27. List of preparers
28. Advance or photocopies of EIS
29. Responses to comments
30. Adoption of EISs
31. Application of regulations to independent regulatory agencies
32. Supplements to old EISs
33. Referrals
34. Records of decision
35. Time required for the NEPA process
36. Environmental assessments (EA)
37. Findings of no significant impact (FONSI)

38. Public availability of EAs vs. FONSIs
39. Mitigation measures imposed in EAs and FONSIs
40. Propriety of issuing EA when mitigation reduces impacts
41. An often-asked question, not covered by the CEQ list of questions: do the purpose and need refer to the NEPA document or to the proposed action? The answer: the purpose and need refer to the proposed action, i.e., what is the purpose of the proposed action and what need does it fulfill? For example, the purpose of constructing the coal-fired power plant, of our earlier example, is to provide the needed electrical power. The need for power has to be discussed in detail relative to available power and projected demands for power over the duration of the operation of the proposed power plant.

II. QUESTIONS AND ANSWERS ABOUT THE NEPA REGULATIONS

1a. **Q.** What is meant by ''range of alternatives'' as referred to in Section 1505.1(e)?

A. The phrase ''range of alternatives'' refers to the alternatives discussed in environmental documents. It includes all reasonable alternatives which must be rigorously explored and objectively evaluated, as well as those other alternatives which are eliminated from detailed study with a brief discussion of the reasons for eliminating them (Section 1502.14). A decision maker must not consider alternatives beyond the range of alternatives discussed in the relevant environmental documents. Moreover, a decision maker must, in fact, consider all the alternatives discussed in an EIS (Section 1505.1(e)).

1b. **Q.** How many alternatives have to be discussed when there is an infinite number of possible alternatives?

A. For some proposals there may exist a very large or even an infinite number of possible reasonable alternatives. For example, a proposal to designate wilderness areas within a national forest could be said to involve an infinite number of alternatives from 0 to 100% of the forest. When there are potentially a very large number of alternatives, only a reasonable number of examples, covering the full spectrum of alternatives, must be analyzed and compared in the EIS. An appropriate series of alternatives might include dedicating 0, 10, 30, 50, 70, 90, or 100% of the forest to wilderness. What constitutes a reasonable range of alternatives depends on the nature of the proposal and the facts in each case.

2a. **Q.** If an EIS is prepared in connection with an application for a permit or other federal approval, must the EIS rigorously analyze and discuss alternatives that are outside the capability of the applicant, or can it be limited to reasonable alternatives that can be carried out by the applicant?

A. Section 1502.14 requires the EIS to examine all reasonable alternatives to the proposal. In determining the scope of alternatives to be considered, the emphasis is on what is ''reasonable'' rather than on whether the proponent or applicant likes it or is itself capable of carrying out a particular alternative. Reasonable alternatives include those that are practical or feasible from the technical and economic standpoint, using common sense, rather than those simply desirable from the standpoint of the applicant.

2b. **Q.** Must the EIS analyze alternatives outside the jurisdiction or capability of the agency or beyond what Congress has authorized?

A. An alternative that is outside the legal jurisdiction of the lead agency must still be analyzed in the EIS if it is reasonable. A potential conflict with local or federal law does not necessarily render an alternative unreasonable, although such conflicts must be considered (Section 1506.2(d)). Alternatives that are outside the scope of what Congress has approved or funded must still be evaluated in the EIS if they are reasonable, because the EIS may serve as the basis for modifying the Congressional approval or funding in light of the goals and policies of NEPA (Section 1500.1(a)).

3. **Q.** What does the "no action" alternative include? If an agency is under a court order or legislative command to act, must the EIS address the "no action" alternative?

A. Section 1502.14(d) requires the alternatives analysis in the EIS to "include the alternative of no action." There are two distinct interpretations of "no action" that must be considered depending on the nature of the proposal being evaluated. The first situation might involve an action such as updating a land management plan where ongoing programs initiated under existing legislation and regulations will continue, even as new plans are developed. In these cases, "no action" is "no change" from current management direction or level of management intensity. To construct an alternative that is based on no management at all would be a useless academic exercise. Therefore, the "no action" alternative may be thought of in terms of continuing with the present course of action until that action is changed. Consequently, projected impacts of alternative management schemes would be compared in the EIS to those impacts projected for the existing plan. In this case, alternatives would include management plans of both greater and lesser intensity, especially greater and lesser levels of resource development.

The second interpretation of "no action" is illustrated in instances involving federal decisions on proposals for projects. "No action" in such cases would mean the proposed activity would not take place, and the resulting environmental effects from taking no action would be compared with the effects of permitting the proposed activity or an alternative activity to go forward.

Where a choice of "no action" by the agency would result in predictable actions by others, this consequence of the "no action" alternative should be included in the analysis. For example, if denial of permission to build a railroad to a facility would lead to construction of a road and increased truck traffic, the EIS should analyze this consequence of the "no action" alternative.

In light of the above, it is difficult to think of a situation where it would not be appropriate to address a "no action" alternative. Accordingly, the regulations require the analysis of the "no action" alternative even if the agency is under a court order or legislative command to act. This analysis provides a benchmark, enabling decision makers to compare the magnitude of environmental effects of the "no action" alternatives. It is also an example of a reasonable alternative outside the jurisdiction of the agency which must be analyzed (Section 2502.14(c); see Question 2 above). Inclusion of such an analysis in the EIS is necessary to inform the Congress, the public, and the president as intended by NEPA (Section 1500.1(a)).

4a. **Q.** What is the "agency's preferred alternative"?

A. The "agency's preferred alternative" is the alternative which the agency believes would fulfill its statutory mission and responsibilities, giving consideration to economic, environmental, technical, and other factors. The concept of the "agency's preferred alternative" is different from the "environmentally preferable alternative", although, in some cases, one alternative may be both (see Question 6 below). It is identified so that agencies and the public can understand the orientation of the lead agency.

4b. **Q.** Does the "preferred alternative" have to be identified in the draft EIS and the final EIS or just in the final EIS?

A. Section 1502.14(e) requires the section of the EIS on alternatives to "identify the agency's preferred alternative if one or more exists, in the draft statement, and identify such alternative in the final statement . . . " This means that if the agency has a preferred alternative at the draft EIS stage, that alternative must be labeled or identified as such in the draft EIS. If the responsible federal official in fact has no preferred alternative at the draft EIS stage, a preferred alternative need not be identified there. By the time the final EIS is filed, Section 1502.14(e) presumes the existence of a preferred alternative and requires its identification in the final EIS, "unless another law prohibits the expression of such a preference."

4c. **Q.** Who recommends or determines the "preferred alternative"?

A. The lead agency official with line responsibility for preparing the EIS and assuring its adequacy is responsible for identifying the preferred alternative(s) of the agency. The NEPA regulations do not dictate which official in an agency shall be responsible for preparation of EISs, but agencies can identify this official in this implementing procedures, pursuant to Section 1507.3.

Even though the preferred alternative of the agency is identified by the EIS preparer in the EIS, the statement must be objectively prepared and not slanted to support the choice of the preferred alternative of the agency over the other reasonable and feasible alternatives.

5a. **Q.** Is the "proposed action" the same thing as the "preferred alternative"?

A. The "proposed action" may be, but is not necessarily, the "preferred alternative" of the agency. The proposed action may be a proposal in its initial form before undergoing anaysis in the EIS process. If the proposed action is internally generated, such as preparing a land management plan, the proposed action might end up as the preferred alternative of the agency. On the other hand, the proposed action may be granting an application to a nonfederal entity for a permit. The agency may or may not have a "preferred alternative" at the draft EIS stage (see Question 4 above). In that case, the agency may decide at the final EIS stage, on the basis of the draft EIS and the public and agency comments, that an alternative other than the proposed action is the "preferred alternative" of the agency.

5b. **Q.** Is the analysis of the "proposed action" in an EIS to be treated differently from the analysis of alternatives?

A. The degree of analysis devoted to each alternative in the EIS is to be substantially similar to that devoted to the "proposed action". Section 1501.3 is titled, "Alternatives included in the proposed action,'' to reflect such comparable treatment. Section 1502.14(9b) specifically requires ''substantial treatment'' in the EIS of each alternative, including the proposed action. This regulation does not dictate an amount of information to be provided, but rather prescribes a level of treatment which may, in turn, require varying amounts of information to enable a reviewer to evaluate and compare alternatives.

6a. **Q.** What is the meaning of the term ''environmentally preferable alternative'', as used in the regulations with reference to records of decision? How is the term ''environment'' used in the phrase?

A. Section 1505.2(b) requires that, in cases where an EIS has been prepared, the record of decision (ROD) must identify all alternatives that were considered, '' . . . specifying the alternative or alternatives which were considered to be environmentally preferable.'' The environmentally preferable alternative is the alternative that will promote the national environmental policy as expressed in NEPA, Section 101. Ordinarily, this means the alternative that causes the least damage to the biological and physical environment; it also means the alternative which best protects, preserves, and enhances historic, cultural, and natural resources.

The Council recognizes that the identification of the environmentally preferable alternative may involve difficult judgments, particularly when one environmental value must be balanced against another. The public and other agencies reviewing a draft EIS can assist the lead agency to develop and determine environmentally preferable alternatives by providing their views in comments on the draft EIS. Through the identification of the environmentally preferable alternative, the decision maker is clearly faced with a choice between that alternative and others, and must consider whether the decision accords with the Congressionally declared policies of the act.

6b. **Q.** Who recommends or determines that which is environmentally preferable?

A. The agency EIS staff is encouraged to make recommendations of the environmentally preferable alternative(s) during EIS preparation. In any event, the lead agency official responsible for the EIS is encouraged to identify the environmentally preferable alternative(s) in the EIS. In all cases, comments from other agencies and the public are also encouraged to address this question. The agency must identify the environmentally preferable alternative in the ROD.

7. **Q.** What is the difference between the sections in the EIS on ''alternatives'' and ''environmental consequences''? How do you avoid duplicating the discussion of alternatives in preparing these two sections?

A. The ''alternatives'' section is the heart of the EIS. This section rigorously explores and objectively evaluates all reasonable alternatives, including the proposed action (Section 1502.14). It should include relevant comparisons on environmental and other grounds. The ''environmental consequences'' section of the EIS discusses the specific environmental impacts or effects of each of the alternatives, including the proposed action (Section 1502.16). In order to avoid duplication between these two sections, most of the ''alternatives'' section should be devoted to describing and comparing the alternatives. Discussion of the environmental impacts of these alternatives should be limited to a concise, descriptive summary of such impacts in a comparative form, including charts or tables, thus sharply defining the issues and providing a clear basis for choice among options (Section 1502.14). The "environmental consequences" section should be devoted largely to a scientific analysis of the direct and indirect environmental effects of the proposed action and of each of the alternatives. It forms the analytic basis for the concise comparison in the "alternatives" section.

8. **Q.** Section 1501.2(d) of the NEPA regulations requires agencies to provide for the early application of NEPA to cases where actions are planned by private applicants or nonfederal entities and are, at some stage, subject to federal approval of permits, loans, loan guarantees, insurance, or other actions. What must and can agencies do to apply NEPA early in these cases?

A. Section 1501.2(d) requires federal agencies to take steps toward ensuring that private parties and state and local entities initiate environmental studies as soon as federal involvement in these proposals can be foreseen. This section is intended to ensure that environmental factors are considered at an early stage in the planning process and to avoid the situation where the applicant for a federal permit or approval has completed planning and eliminated all alternatives to the proposed action by the time the EIS process commences or before the EIS process has been completed.

Through early consultation, business applicants and approving agencies may gain better appreciation of each other's needs and foster a decision making process which avoids later unexpected confrontations.

Federal agencies are required by Section 1507.3(b) to develop procedures to carry out Section 1501.2(d). The procedures should include an "outreach program", such as a means for prospective applicants to conduct preapplication consultations with the lead and cooperating agencies. Applicants need to find out in advance of project planning what environmental studies or other information will be required and what mitigation requirements are likely in connection with the later federal NEPA process. Agencies should designate staff to advise potential applicants of the NEPA information requirements of the agency and should publicize their preapplication procedures and information requirements in newsletters or other media used by potential applicants.

Complementing Section 1501.2(d), Section 1506.5(a) requires agencies to assist applicants by outlining the types of information required in those cases where the agency requires the applicant to submit environmental data for possible use by the agency in preparing an EIS.

Section 1506.5(b) allows agencies to authorize preparation of environmental assessments by applicants. Thus, the procedures should also include a means for anticipating and utilizing applicants' environmental studies or "early corporate environmental assessments" to fulfill some of the NEPA obligations of the federal agency. However, in such cases, the agency must still evaluate independently the environmental issues and take responsibility for the environmental assessment.

These provisions are intended to encourage and enable private and other nonfederal entities to build environmental considerations into their own planning processes in a way that facilitates the application of NEPA and avoids delay.

9. **Q.** To what extent must an agency inquire into whether an applicant for a federal permit, funding, or other approval of a proposal will also need approval from another agency for the same proposal or some other related aspect of it?

A. Agencies must integrate the NEPA process into other planning at the earliest possible time to insure that planning and decisions reflect environmental values, to avoid delays later in the process, and to head off potential conflicts. Specifically, the agency must "provide for cases where actions are planned by . . . applications," so that designated staff are available to advise potential applicants of studies or other information that will foreseeably be required for the later federal action; the agency shall consult with the applicant if the agency foresees its own involvement in the proposal; and it shall insure that the NEPA process commences at the earliest possible time (Section 1501.2(d); see Question 8).

The regulations emphasize agency cooperation early in the NEPA process (Section 1501.5). Section 1501.7 on "scoping" also provides that all affected federal agencies are to be invited to participate in scoping the environmental issues and to identify the various environmental review and consultation requirements that may apply to the proposed action. Further, Section 1502.25(b) requires that the draft EIS list all the federal permits, licenses, and other entitlements that are needed to implement the proposal.

These provisions create an affirmative obligation on federal agencies to inquire early and, to the maximum degree possible, to ascertain whether an applicant is or will be seeking other federal assistance or approval, or whether the applicant is waiting until a proposal has been substantially developed before requesting federal aid or approval.

Thus, a federal agency receiving a request for approval or assistance should determine whether the applicant has filed separate requests for federal approval or assistance with other federal agencies. Other federal agencies that are likely to become involved should then be contacted, and the NEPA process coordinated to insure an early and comprehensive analysis of the direct and indirect effects of the proposal and any related actions. The agency should inform the applicant the action on its application may be delayed unless it submits all other federal applications (where feasible to do so) so that all the relevant agencies can work together on the scoping process and preparation of the EIS.

10a. **Q.** What actions by agencies and/or applicants are allowed during EIS preparation and during the 30-day review period after publication of a final EIS?

A. No federal decision on the proposed action shall be made or recorded until at least 30 days after the publication by EPA of notice that the particular EIS has been filed with EPA (Sections 1505.2 and 1506.10). Section 1505.2 requires this decision to be stated in a public record of decision. Until the agency issues its record of decision, no action by an agency or applicant concerning the proposal shall be taken which would have an adverse environmental impact or limit the choice of reasonable alternatives (Section 1506.1(a)). But this does not preclude preliminary planning or design work which is needed to support an application for permits or assistance (Section 1506.1(d).

When the impact statement in question is a program EIS, no major action concerning the program may be taken which may significantly affect the quality of the human environment, unless the particular action is justified independently of the program, is accompanied by its own adequate environmental impact statement, and will not prejudice the ultimate decision on the program (Section 1506.1(c)).

10b. **Q.** Do these limitations on action (described in Question 10a) apply to state or local agencies that have statutorily delegated responsibility for preparation of environmental documents required by NEPA, for example, under the HUD Block Grant program?

A. Yes, these limitations do apply, without any variation from their application to federal agencies.

11. **Q.** What actions must a lead agency taken during the NEPA process when it becomes aware that a nonfederal applicant is about to take an action within the jurisdiction of the agency that would either have an adverse environmental impact or limit the choice of reasonable alternatives (e.g., prematurely commit money or other resources towards the completion of the proposal)?

A. The federal agency must notify the applicant that the agency will take strong affirmative steps to insure that the objectives and procedures of NEPA are fulfilled (Section 1506.1(b). These steps could include seeking injunctive measurements under NEPA, or the use of sanctions available under either the permitting authority of the agency or statutes setting forth the statutory mission of the agency. For example, the agency might advise an applicant that if such action is taken, the agency will not process the application.

12a. **Q.** What actions are subject to the new regulations of CEQ, and what actions are grandfathered under the old guidelines?

A. The effective date of the CEQ regulations was July 30, 1979 (except for certain HUD programs under the Housing and Community Development Act, 42 U.S.C. 5304(h), and certain state highway programs that qualify under Section 102(2)(D) of NEPA for which the regulations became effective on November 30, 1979). All the provisions of the regulations are binding as of that date, including those covering decision making, public participation, referrals, limitations on actions, EIS supplements, etc. For example, a record of decision would be prepared even for decisions where the draft EIS was filed before July 30, 1979.

But in determining whether or not the new regulations apply to the preparation of a particular environmental document, the relevant factor is the date of filing of the draft of that document. Thus, the new regulations do not require the redrafting of an EIS or supplement if the draft EIS or supplement was filed before July 30, 1979. However, a supplement prepared after the effective date of the regulations for an EIS issued in final before the effective date of the regulations would be controlled by the regulations.

Even though agencies are not required to apply the regulations to an EIS or other document for which the draft was filed prior to July 30, 1979, the regulations encourage agencies to follow the regulations "to the fullest extent practicable," i.e., if it is feasible to do so, in preparing the final document (Section 1506.12(a)).

12b. **Q.** Are projects authorized by Congress before the effective date of the CEQ regulations grandfathered?

A. No. The date of Congressional authorization for a project is not determinative of whether the CEQ regulations or former guidelines apply to the particular proposal. No incomplete projects or proposals of any kind are grandfathered in whole or in part. Only certain environmental documents, for which the draft was issued before the effective date of the regulations, are grandfathered and subject to the former guidelines of the CEQ.

12c. **Q.** Can a violation of the regulations give rise to a cause of action?

A. While a trivial violation of the regulations would not give rise to an independent cause of action, such a cause of action would arise from a substantial violation of the regulations (Section 1500.3).

13. **Q.** Can the scoping process be used in connection with preparation of an environmental assessment, i.e., before both the decision to proceed with an EIS and publication of a notice of intent?

A. Yes. Scoping can be a useful tool for discovering alternatives to a proposal or significant impacts that may have been overlooked. In cases where an environmental assessment is being prepared to help an agency decide whether to prepare an EIS, useful information might result from early participation by other agencies and the public in a scoping process.

The regulations state that the scoping process is to be preceded by a notice of intent (NOI) to prepare an EIS. But that is only the minimum requirement. Scoping may be initiated earlier, as long as there is appropriate public notice and enough information available on the proposal so that the public and relevant agencies can participate effectively.

However, scoping that is done before the assessment, and in aid of its preparation, cannot substitute for the normal scoping process after publication of the NOI, unless the earlier public notice stated clearly that this possibility was under consideration, and the NOI expressly provides that written comments on the scope of alternatives and impacts will still be considered.

14a. **Q.** What are the respective rights and responsiblities of lead and cooperating agencies? What letters and memoranda must be prepared?

A. After a lead agency has been designated (Section 1501.5), that agency has the responsibility to solicit cooperation from other federal agencies that have jurisdiction by law or special expertise on any environmental issue that should be addressed in the EIS being prepared. Where appropriate, the lead agency should seek the cooperation of state or local agencies of similar qualifications. When the proposal may affect an Indian reservation, the agency should consult with the Indian tribe (Section 1508.5). The request for cooperation should come at the earliest possible time in the NEPA process.

After discussions with the candidate cooperating agencies, the lead agency and the cooperating agencies are to determine by letter or by memorandum which agencies will undertake cooperating responsibilities. To the extent possible at this stage, responsibilities for specific issues should be assigned. The allocation of responsibilities will be completed during scoping (Section 1501.7(a)(4)).

Cooperating agencies must assume responsibility for the development of information and the preparation of environmental analyses at the request of the lead agency (Section 1501.6(b)(3)). Cooperating agencies are not required by Section 1501.6 to devote staff resources that were normally used to critique or comment on the draft EIS after its preparation, much earlier in the NEPA process — primarily at the scoping and draft EIS preparation stages. If a cooperating agency determines that its resource limitations preclude any involvement, or the degree of involvement (amount of work) requested by the lead agency, it must so inform the lead agency in writing and submit a copy of this correspondence to the Council (Section 1501.6(c)).

In other words, the potential cooperating agency must decide early if it is able to devote any of its resources to a particular proposal. For this reason, the regulation states that an agency may reply to a request for cooperation that "other program commitments preclude any involvement or the degree of involvement requested in the action that is the subject of the environmental impact statement." (Emphasis added.) The regulations refers to the "action", rather than to the EIS, to clarify that the agency is taking itself out of all phases of the federal action, not just draft EIS preparation. This means that the agency has determined that it cannot be involved in the later stages of EIS review and comment, as well as decision making on the proposed action. For this reason, cooperating agencies with jurisdiction by law (those which have permitting or other approval authority) cannot opt out entirely of the duty to cooperate on the EIS (see also Question 15, relating specifically to the responsibility of EPA).

14b. **Q.** How are disputes resolved between lead and cooperating agencies concerning the scope and level of detail of analysis and the quality of data in impact statements?

A. Such disputes are resolved by the agencies themselves. A lead agency, of course, has the ultimate responsibility for the content of an EIS. But it is supposed to use the environmental analysis and recommendations of cooperating agencies with jurisdiction by law or special expertise to the maximum extent possible, consistent with its own responsibilities as lead agency (Section 1501.6(a)(2)).

If the lead agency leaves out a significant issue or ignores the advice and expertise of the cooperating agency, the EIS may be found later to be inadequate. Similarly, where cooperating agencies have their own decisions to make and they intend to adopt the environmental impact statement and base their decisions on it, one document should include all of the information necessary for the decisions by the cooperating agencies. Otherwise they may be forced to duplicate the EIS process by issuing a new, more complete EIS or supplement EIS, even though the original EIS could have sufficed if it had been properly done at the outset. Thus, both lead and cooperating agencies have a stake in producing a document of good quality. Cooperating agencies also have a duty to participate fully in the scoping process to ensure that the appropriate range of issues is determined early in the EIS process.

Because the EIS is not the record of decision, but instead constitutes the information and analysis on which to base a decision, disagreements about conclusions to be drawn from the EIS need not inhibit agencies from issuing a joint document, or adopting an EIS of another agency, if the analysis is adequate. Thus, if each agency has its own "preferred alternative", both can be identified in the EIS. Similarly, a cooperating agency with jurisdiction by law may determine in its own ROD that Alternative A is the environmentally preferable action, even though the lead agency has decided in its separate ROD that Alternative B is environmentally preferable.

14c. **Q.** What are the specific responsibilities of federal and state cooperating agencies to review draft EISs?

A. Cooperating agencies (i.e., agencies with jurisdiction by law or special expertise) and agencies that are authorized to develop or enforce environmental standards must comment on environmental impact statements within their jurisdiction, expertise, or authority (Sections 1503.2, 1508.5). If a cooperating agency is satisfied that its views are adequately reflected in the environmental impact statement, it should simply comment accordingly. Conversely, if the cooperating agency determines that a draft EIS is incomplete, inadequate, or inaccurate, or it has other comments, it should promptly make such comments, conforming to the requirements of specificity in Section 1503.3.

14d. **Q.** How is the lead agency to treat the comments of another agency with jurisdiction by law or special expertise which has failed or refused to cooperate or participate in scoping or EIS preparation?

A. A lead agency has the responsibility to respond to all substantive comments raising significant issues regarding a draft EIS (Section 1500.4). However, cooperating agencies are generally under an obligation to raise issues or otherwise participate in the EIS process during scoping and EIS preparation if they reasonably can do so. In practical terms, if a cooperating agency fails to cooperate at the outset, such as during scoping, it will find that its comments at a later stage will not be as persuasive to the lead agency.

15. **Q.** Are responsibilities of the EPA to review and comment on the environmental effects of agency proposals under Section 309 of the Clean Air Act independent of its responsiblity as a cooperating agency?

A. Yes. EPA has an obligation under Section 309 of the Clean Air Act to review and comment in writing on the environmental impact of any matter relating to the authority of the administrator contained in proposed legislation, federal construction projects, other federal actions requiring EISs, and new regulations (42 U.S.C. Sec. 7609). This obligation is independent of its role as a cooperating agency under the NEPA regulations.

16. **Q.** What is meant by the term "third party contracts" in connection with the preparation of an EIS? (See Section 1506.5(c).) When can "third party contracts" be used?

A. As used by EPA and other agencies, the term "third party contract" refers to the preparation of EISs by contractors paid by the applicant. In the case of an EIS for a National Pollution Discharge Elimination System (NPDES) permit, the applicant, aware in the early planning stages of the proposed project of the need for an EIS, contracts directly with a consulting firm for its preparation (Section 40 CFR 6.604(g)). The "third party" is EPA which, under Section 1506.5(c) must select the consulting firm, even though the applicant pays for the cost of preparing the EIS. The consulting firm is responsible to EPA for preparing an EIS that meets the requirements of the NEPA regulations and the NEPA procedures of the EPA. It is in the applicant's interest that the EIS can comply with the law so that EPA can take prompt action on the NPDES permit application. The "third party contract" method under the NEPA procedures of the EPA is purely voluntary, though most applicants have found it helpful in expediting compliance with NEPA.

If a federal agency used "third party contracting", the applicant may undertake the necessary paperwork for the solicitation of a field of candidates under the direction of the agency, so long as the agency complies with Section 1506.5(c). Federal procurement requirements do not apply to the agency because it incurs no obligations or costs under the contract, nor does the agency procure anything under the contract.

17a. **Q.** If an EIS is prepared with the assistance of a consulting firm, the firm must execute a disclosure statement. What criteria must the firm follow in determining whether it has any "financial or other interest in the outcome of the project" which would cause a conflict of interest?

A. Section 1506.5(c), which specifies that a consulting firm preparing an EIS must execute a disclosure statement, does not define "financial or other interest in the outcome of the project." The Council interprets this term broadly to cover any known benefits other than general enhancement of professional reputation. This includes any financial benefit such as a promise of future construction or design work on the project as well as indirect benefits the consultant is aware of (e.g., if the project would aid proposals sponsored by the other clients of the firm). For example, completion of a highway project may encourage construction of a shopping center or industrial park from which the consultant stands to benefit. If a consulting firm is aware that it has such an interest in the decision on the proposal, it should be disqualified from preparing the EIS to preserve the objectivity and integrity of the NEPA process.

When a consulting firm has been involved in developing initial data and plans for the project, but does not have any financial or other interest in the outcome of the decision, it need not be disqualified from preparing the EIS. However, a disclosure statement in the draft EIS should clearly state the scope and extent of the prior involvement of the firm to expose any potential conflicts of interest that may exist.

17b. **Q.** If the firm in fact has no promise of future work or other interest in the outcome of the proposal, may the firm later bid in competition with others for future work on the project if the proposed action is approved?

A. Yes.

18. **Q.** How should uncertainties about indirect effects of a proposal be addressed, for example, in cases of disposal of federal lands when the identity or plans of future landowners is unknown?

A. The EIS must identify all the indirect effects that are known, and make a good faith effort to explain the effects that are not known, but are "reasonably foreseeable" (Section 1508.8(b)). In the example, if there is total uncertainty about the identity of future land owners or the nature of future land uses, then of course the agency is not required to engage in speculation or contemplation about their future plans. But, in the ordinary course of business, people do make judgments based upon reasonably foreseeable occurrences. It will often be possible to consider the likely purchasers and the development trends in that area or similar areas in recent years; or the likelihood that the land will be used for an energy project, shopping center, subdivision, farm, or factory. The agency has the responsibility to make an informed judgment and to estimate future impacts on that basis, especially if trends are ascertainable or potential purchasers have made themselves known. The agency cannot ignore these uncertain, but probable, effects of its decisions.

19a. **Q.** What is the scope of mitigation measures that must be discussed?

A. The mitigation measures discussed in an EIS must cover the range of impacts on the proposal. The measures must include such things as design alternatives that would decrease pollution emissions, construction impacts, esthetic intrusion, as well as relocation assistance, possible land use controls that could be enacted, and other possible efforts. Mitigation measures must be considered even for impacts that by themselves would not be considered "significant". Once the proposal itself is considered as a whole to have significant effects, all of its specific effects on the environment (whether or not "significant") must be considered, and mitigation measures must be developed where it is feasible to do so (Sections 1502.14(f), 1502.16(h), 1508.14).

19b. **Q.** How should an EIS treat the subject of available mitigation measures that are (1) outside the jurisdiction of the lead agency or cooperating agencies or (2) unlikely to be adopted or enforced by the responsible agency?

A. All relevant, reasonable mitigation measures that could improve the project are to be identified, even if they are outside the jurisdiction of the lead agency or the cooperating agencies and thus would not be committed as part of the RODs of these agencies (Sections 1502.16(h), 1505.2(c)). This will serve to alert agencies or officials who can implement these extra measures and will encourage them to do so. Because the EIS is the most comprehensive environmental document, it is an ideal vehicle in which to lay out not only the full range of environmental impacts, but also the full spectrum of appropriate mitigation.

However, to ensure that environmental effects of a proposed action are fairly assessed, the probability of the mitigation measures being implemented must also be discussed. Thus, the EIS and the record of decision should indicate the likelihood that such measures will be adopted or enforced by the responsible agencies (Sections 1502.16(h), 1505.2). If there is a history of nonenforcement or opposition to such measures, the EIS and record of decision should acknowledge such opposition or nonenforcement. If the necessary mitigation measures will not be ready for a long period of time, this fact, of course, should also be recognized.

20a. **Q.** When must a worst case analysis be included in an EIS?

A. If there are gaps in relevant information or scientific uncertainty pertaining to an evaluation of an agency of significant adverse impacts on the human environment, an agency must make clear that such information is lacking or that the uncertainty exists. An agency must include a worst case analysis of the potential impacts of the proposal and an indication of the probability or improbability of their occurrence if (a) the information relevant to adverse impacts is essential to a reasoned choice among alternatives and the overall costs of obtaining the information are exorbitant, or (b) the information relevant to adverse impacts is important to the decision and the means to obtain it are not known. NEPA requires that impact statements, at a minimum, contain information to alert the public and Congress to all known possible environmental consequences of agency action. Thus, one of the most important obligations of the federal government is to present to the fullest extent possible the spectrum of consequences that may result from agency decisions and the details of their potential consequences for the human environment.

20b. **Q.** What is the purpose of a worst case analysis? How is it formulated, and what is the scope of the analysis?

A. The purpose of the analysis is to carry out the mandate of NEPA for full disclosure to the public of the potential consequences of any agency decisions, and to cause agencies to consider those potential consequences when acting on the basis of scientific uncertainties or gaps in available information. The analysis is formulated on the basis of available information, using reasonable projections of the worst possible consequences of a proposed action.

For example, if there are scientific uncertainties and gaps in the available information concerning the numbers of juvenile fish that would be entrained in a cooling water facility, the responsible agency must disclose and consider the possibility of the loss of the commercial or sport fishery.

In addition to an analysis of a low probability/catastrophic impact event, the worst case analysis should also include a spectrum of events of higher probability, but of less drastic impact.

21. **Q.** Where an EIS or an EA is combined with another project planning document (sometimes called ''piggybacking''), to what degree may the EIS or EA refer to and rely upon information in the project document to satisfy the requirements of NEPA?

A. Section 1502.25 of the regulations requires that draft EISs be prepared concurrently and integrated with environmental analyses and related surveys and studies required by other federal statutes. In addition, Section 1506.4 allows any environmental document prepared in compliance with NEPA to be combined with any other agency document to reduce duplication and paperwork. However, these provisions were not intended to authorize the preparation of a short summary or outline EIS, attached to a detailed project report or land use plan containing the required environmental impact data. In such circumstances, the reader would have to refer constantly to the detailed report to understand the environmental impacts and alternatives which would have been found in the EIS itself.

The EIS must stand on its own as an analytical document which fully informs decision makers and the public of the environmental effects of the proposal and those of the reasonable alternatives (Secton 1502.1). But, as long as the EIS is clearly identified and is self-supporting, it can be physically included in or attached to the project report or land use plan and may use attached report material as technical backup.

Forest Service environmental impact statements for forest management plans are handled in this manner. The EIS identifies the preferred alternative of the agency, which is developed in detail as the proposed management plan. The detailed proposed plan accompanies the EIS through the review process, and documents are appropriately cross referenced. The proposed plan is useful for EIS readers as an example to show how one choice of management options translates into effects on natural resources. This procedure permits initiation of the 90-day public review of proposed forest plans which is required by the National Forest Management Act.

All the alternatives are discussed in the EIS which can be read as an independent document. The details of the management plan are not repeated in the EIS, and vice versa. This is a reasonable functional separation of the documents: the EIS contains information relevant to the choice among alternatives; the plan is a detailed description of proposed management activities suitable for use by the land managers. This procedure provides for concurrent compliance with the public review requirements of both NEPA and the National Forest Management Act.

Under some circumstances, a project report or management plan may be totally merged with the EIS, and the one document labeled as both ''EIS'' and ''management plan'' or ''project report''. This may be reasonable where the documents are short or where the EIS format and the regulations for clear, analytical EISs also satisfy the requirements for a project report.

22. **Q.** May state and federal agencies serve as joint lead agencies? If so, how do they resolve law, policy, and resource conflicts under NEPA and the relevant state environmental policy act? How do they resolve differences in perspective where, for example, national and local needs may differ?

A. Under Section 1501.5(b), federal, state, or local agencies, as long as they include at least one federal agency, may act as joint lead agencies to prepare an EIS. Section 1506.2 also strongly urges state and local agencies and the relevant federal agencies to cooperate fully with each other. This should cover joint research and studies, planning activities, public hearings, environmental assessments, and the preparation of joint EISs under NEPA and the relevant ''little NEPA'' state laws so that one document will satisfy both laws.

The regulations also recognize that certain inconsistencies may exist between the proposed federal action and any approved state or local plan or law. The joint document should discuss the extent to which the federal agency would reconcile its proposed action with such plan or law (Section 1506.2(d); see Question 23).

Because there may be differences in perspective, as well as conflicts among federal, state, and local goals for resources, management, the Council has advised participating agencies to adopt a flexible, cooperative approach. The final EIS should reflect all of their interests and missions, clearly identified as such. The final document would then indicate how state and local interest have been accommodated, or would identify conflicts in goals (e.g., how a hydroelectric project, which might induce second home development, would require new land use controls). The EIS must contain a complete discussion of scope and purpose of the proposal, alternatives, and impacts so that the discussion is adequate to meet the needs of local, state, and federal decision makers.

23a. **Q.** How should an agency handle potential conflicts between a proposal and the objectives of federal, state, or local land use plans, policies, and controls for the area concerned? (See Section 1502.10(c)).

A. The agency should first inquire of other agencies whether there are any potential conflicts. If there would be immediate conflicts, or if conflicts could arise in the future when the plans are finished (see Question 23b below), the EIS must acknowledge and describe the extent of those conflicts. If there are any possibilities of resolving the conflicts, these should be explained as well. The EIS should also evaluate the seriousness of the impact of the proposal on the land use plans and policies and whether, or how much, the proposal will impair the effectiveness of land use control mechanisms for the area. Comments from officials of the affected area should be solicited early and should be carefully acknowledged and answered in the EIS.

23b. **Q.** What constitutes a ''land use plan or policy'' for purposes of this discussion?

A. The term ''land use plans'', includes all types of formally adopted documents for land use planning, zoning, and related regulatory requirements. Local general plans are included, even though they are subject to future change. Proposed plans should also be addressed if they have been formally proposed by the appropriate government body in a written form and are being actively pursued by officials of the jurisdiction. Staged plans, which must go through phases of development such as the Level A, B, and C planning process of the Water Resources Council, should also be included, even though they are incomplete.

The term "policies" includes formally adopted statements of land use policy as embodied in laws or regulations. It also includes proposals for action such as the initiation of a planning process or a formally adopted policy statement of the local, regional, or state executive branch, even if it has not yet been formally adopted by the local, regional, or state legislative body.

23c. **Q.** What options are available for the decision maker when conflicts with such plans or policies are identified?

A. After identifying any potential land use conflicts, the decision maker must weigh the significance of the conflicts among all the other environmental and nonenvironmental factors that must be considered in reaching a rational and balanced decision. Unless precluded by other law from causing or contributing to any inconsistency with the land use plans, policies, or controls, the decision maker retains the authority to go forward with the proposal, despite the potential conflict. In the record of decision, the decision maker must explain what the decision was, how it was made, and what mitigation measures are being imposed to lessen adverse environmental impacts of the proposal, among the other requirements of Section 1505.2. This provision would require the decision maker to explain any decision to override land use plans, policies, or controls for the area.

24a. **Q.** When are EISs required on policies, plans or programs?

A. An EIS must be prepared if an agency proposes to implement a specific policy, to adopt a plan for a group of related actions, or to implement a specific statutory program or executive directive (Section 1508.18). In addition, the adoption of official policy in the form of rules, regulations, and interpretations pursuant to the Administrative Procedure Act, treaties, conventions, or other formal documents establishing governmental or agency policy which will substantially alter agency programs could require an EIS (Section 1508.18). In all cases, the policy, plan, or program must have the potential for significantly affecting the quality of the human environment in order to require an EIS. It should be noted that a proposal "may exist in fact as well as by agency declaration that one exists" (Section 1508.23).

24b. **Q.** When is an area-wide or overview EIS appropriate?

A. The preparation of an area-wide or overview EIS may be particularly useful when similar actions, viewed with other reasonably foreseeable or proposed agency actions, share common timing or geography. For example, when a variety of energy projects may be located in a single watershed, or when a series of new energy technologies may be developed through federal funding, the overview or area-wide EIS would serve as a valuable and necessary analysis of the affected environment and the potential cumulative impacts of the reasonably foreseeable actions under that program or within that geographical area.

24c. **Q.** What is the function of tiering in such cases?

A. Tiering is a procedure which allows an agency to avoid duplication of paperwork through the incorporation by reference of the general discussions and relevant specific discussions from an environmental impact statement of broader scope into one of lesser scope or vice versa. In the example given in Question 24b, this would mean that an overview EIS would be prepared for all of the energy activities reasonably foreseeable in a particular geographic area or resulting from a particular development program. This impact statement would be followed by site- or project-specific EISs. The tiering process would make each EIS of greater use and meaning to the public as the plan or program develops, without duplication of the analysis prepared for the previous impact statement.

25a. **Q.** When is it appropriate to use appendices instead of including information in the body of an EIS?

A. The body of the EIS should be a succinct statement of all the information on environmental impacts and alternatives that the decision maker and the public need in order to make the decision and to ascertain that every significant factor has been examined. The EIS must explain or summarize methodologies of research and modeling and the results of research that may have been conducted to analyze impacts and alternatives.

Lengthy technical discussions of modeling methodology, baseline studies, or other work are best reserved for the appendix. In other words, if only technically trained individuals are likely to understand a particular discussion, then it should go in the appendix, and a plain language summary of the analysis and conclusions of that technical discussion should go in the text of the EIS.

The final statement must also contain the response of the agency to comments on the draft EIS. These responses will be primarily in the form of changes in the document itself, but specific answers to each significant comment should also be included. These specific responses may be placed in an appendix. If the comments are especially voluminous, summaries of the comments and responses will suffice. (See Question 29 regarding the level of detail required for responses to comments.)

25b. **Q.** How does an appendix differ from incorporation by reference?

A. First, if at all possible, the appendix accompanies the EIS, whereas the material which is incorporated by reference does not accompany the EIS. Thus, the appendix should contain information that reviewers will likely want to examine. The appendix should include material that pertains to preparation of a particular EIS. Research papers directly relevant to the proposals, lists of affected species, discussion of the methodology of models used in the analysis of impacts, extremely detailed responses to comments, or other information would be placed in the appendix.

The appendix must be complete and available at the time the EIS is filed. Five copies of the appendix must be sent to EPA with five copies of the EIS for filing. If the appendix is too bulky to be circulated, it instead must be placed in conveniently accessible locations or furnished directly to commentors upon request. If it is not circulated with EIS, the Notice of Availability published by EPA must so state, giving a telephone number to enable potential commentors to locate or request copies of the appendix promptly.

Material that is not directly related to preparation of the EIS should be incorporated by reference. This would include other EISs, research papers in the general literature, technical background papers, or other material that someone with technical training could use to evaluate the analysis of the proposal. These must be made available, either by citing the literature, furnishing copies to central locations, or sending copies directly to commentors upon request.

Care must be taken in all cases to ensure that material incorporated by reference, and the occasional appendix that does not accompany the EIS, are in fact available for the full minimum public comment period.

26a. **Q.** How detailed must an EIS index be?

A. The EIS index should have a level of detail sufficient to focus on areas of the EIS of reasonable interest to any reader. It cannot be restricted to the most important topics. On the other hand, it need not identify every conceivable term or phrase in the EIS. If an agency believes that the reader is reasonably likely to be interested in a topic, it should be included.

26b. **Q.** Is a key word index required?

A. No. A key word index is a relatively short list of descriptive terms that identifies the key concepts or subject areas in a document. For example, it could consist of 20 terms which describe the most significant aspects of an EIS that a future researcher would need: type of proposal, type of impacts, type of environment, geographical area, sampling or modelling methodologies used. This technique permits the compilation of EIS data banks, by facilitating quick and inexpensive access to stored materials. While a key word index is not required by the regulations, it could be a useful addition for several reasons. First, it can be useful as a quick index for reviewers of the EIS, helping to focus on areas of interest. Second, if an agency keeps a listing of the key word indexes of the EISs it produces, the EIS preparers themselves will have quick access to similar research data and methodologies to aid their future EIS work. Third, a key word index will be needed to make an EIS available for future researchers using EIS data banks that are being developed. Preparation of such an index now, when the document is produced, will save a later effort when the data banks become operational.

27a. **Q.** If a consultant is used in preparing an EIS, must the list of preparers identify members of the consulting firm as well as the agency NEPA staff who were primarily responsible?

A. Section 1502.17 requires identification of the names and qualifications of persons who were primarily responsible for preparing the EIS or significant background papers, including basic components of the statement. This means that members of a consulting firm preparing material that is to become part of the EIS must be identified. The EIS should identify these individuals even though the consultant's contribution may have been modified by the agency.

27b. **Q.** Should agency staff involved in reviewing and editing the EIS also be included in the list of preparers?

A. Agency personnel who wrote basic components of the EIS or significant background papers must, of course, be identified. The EIS should also list the technical editors who reviewed or edited the statements.

27c. **Q.** How much information should be included on each person listed?

A. The list of preparers should normally not exceed two pages. Therefore, agencies must determine which individuals had primary responsiblity and need not identify individuals with minor involvement. The list of preparers should include a very brief identification of the individuals involved, their qualifications (expertise, professional disciplines), and the specific portion of the EIS for which they are responsible. This may be done in tabular form to cut down on length. A line or two for each person's qualifications should be sufficient.

28. **Q.** May an agency file photocopies of an EIS with EPA pending the completion of printing the document?

A. Photocopies of an EIS may be filed with EPA prior to printing only if the photocopies are simultaneously made available to other agencies and the public. Section 1506.9 of the regulations, which governs EIS filing, specifically requires federal agencies to file EISs with EPA no earlier than the EIS is distributed to the public. However, this section does not prohibit photocopying as a form of reproduction and distribution. When an agency chooses photocopying as the reproduction method, the EIS must be clear and legible to permit ease of reading and ultimate microfiching of the EIS. Where color graphs are important to the EIS, they should be reproduced and circulated with the photocopy.

29a. **Q.** What response must an agency provide to a comment on a draft EIS which states that the EIS methodology is inadequate or inadequately explained? For example, what level of detail must an agency include in its response to a simple post card comment making such an allegation?

A. Appropriate responses to comments are described in Section 1503.4. Normally, the responses should result in changes in the text of the EIS, not simply a separate answer at the back of the document. But, in addition, the agency must state what its response was, and if the agency decides that no substantive response to a comment is necessary, it must briefly explain why.

An agency is not under an obligation to issue a lengthy reiteration of its methodology for any portion of an EIS if the only comment addressing the methodology is a simple complaint that the EIS methodology is inadequate. But agencies must respond to comments, however brief, which are specific in their criticism of agency methodology. For example, if a commentor of an EIS said that the air quality dispersion analysis or methodology of an agency was inadequate, and the agency had included a discussion of that analysis in the EIS, little if anything need be added in response to such a comment. However, if the commentor said that the dispersion analysis was inadequate because of its use of certain computational techniques, or that a dispersion analysis was inadequately explained because computational techniques were not included or referenced, then the agency would have to respond in a substantive and meaningful way to such a comment.

If a number of comments are identical or very similar, agencies may group the comments and prepare a single answer for each group. Comments may be summarized if they are especially voluminous. The comments or summaries must be attached to the EIS, regardless of whether the agency believes they merit individual discussion in the body of the final EIS.

29b. **Q.** How must an agency respond to a comment on a draft EIS that raises a new alternative not previously considered in the draft EIS?

A. This question might arise in several possible situations. First, a commentor on a draft EIS may indicate that there is a possible alternative which, in the view of the agency, is not a reasonable alternative (Section 1502.14(a)). If that is the case, the agency must explain why the comment does not warrant further agency response, citing authorities or reasons that support the position of the agency and, if appropriate, indicate those circumstances which would trigger agency reappraisal or further response (Section 1503.4(a)). For example, a commentor on a draft EIS on a coal-fired power plant may suggest the alternative of using synthetic fuel. The agency may reject the alternative with a brief discussion (with authorities) of the unavailability of synthetic fuel within the time frame necessary to meet the need and purpose of the proposed facility.

A second possibility is that an agency may receive a comment indicating that a particular alternative, while reasonable, should be modified somewhat, for example, to achieve certain mitigation benefits or for other reasons. If the modification is reasonable, the agency should include a discussion of it in the final EIS. For example, a commentor on a draft EIS on a proposal for a pumped storage power facility might suggest that the applicant's proposed alternative should be enhanced by the addition of certain reasonable mitigation measures, including the purchase and setting aside of a wildlife preserve to substitute for the tract to be destroyed by the project. The modified alternative including the additional mitigation measures should be discussed by the agency in the final EIS.

A third, slightly different possibility is that a comment on a draft EIS will raise an alternative which is a minor variation of one of the alternatives discussed in the draft EIS, but this variation was not given any consideration by the agency. In such a case, the agency should develop and evaluate the new alternative, if it is reasonable, in the final EIS. If it is qualitatively within the spectrum of alternatives that were discussed in the draft, a supplemental draft will not be needed. For example, a commentor on a draft EIS to designate a wilderness area within a national forest might reasonably identify a specific tract of the forest and urge that it be considered for designation. If the draft EIS considered designation of a range of alternative tracts which encompassed forest area of similar quality and quantity, no supplemental EIS would have to be prepared. The agency could fulfill its obligation by addressing that specific alternative in the final EIS.

As another example, an EIS on an urban housing project may analyze the alternatives of constructing 2000, 4000, or 6000 units. A commentor on the draft EIS might urge the consideration of constructing 5000 units, utilizing a different configuration of buildings. This alternative is within the spectram of alternatives already considered and, therefore, could be addressed in the final EIS.

A fourth possibility is that a commentor points out an alternative which is not a variation of the proposal or of any alternative discussed in the draft impact statement and is a reasonable alternative that warrants serious agency response. In such a case, the agency must issue a supplement to the draft EIS that discusses this new alternative. For example, a commentor on a draft EIS on a nuclear power plant might suggest that a reasonable alternative for meeting the projected need for power would be through peak load management and energy conservation programs. If the permitting agency has failed to consider that approach in the draft EIS, and the approach cannot be dismissed by the agency as unreasonable, a supplement to the draft EIS which discusses that alternative must be prepared. (If necessary, the same supplement should also discuss substantial changes in the proposed action or significant new circumstances or information, as required by Section 1502.9(c)(1) of the CEQ regulations.

If the new alternative was not raised by the commentor during scoping, but could have been, commentors may find that they are unpersuasive in their efforts to have their suggested alternative analyzed in detail by the agency. However, if the new alternative is discovered or developed later, and it could not reasonably have been raised during the scoping process, then the agency must address it in a supplemental draft EIS. The agency is, in any case, ultimately responsible for preparing an adequate EIS that considers all alternatives.

30. **Q.** When a cooperating agency with jurisdiction by law intends to adopt the EIS of a lead agency and it is not satisfied with the adequacy of the document, may the cooperating agency adopt only the part of the EIS with which it is satisfied? If so, would a cooperating agency with jurisdiction by law have to prepare a separate EIS or EIS supplement covering the areas of disagreement with the lead agency?

A. Generally, a cooperating agency may adopt the EIS of a lead agency without recirculating it if it concludes that its NEPA requirements and its comments and suggestions have been satisfied (Section 1506.3(a),(c)). If necessary, a cooperating agency may adopt only a portion of the EIS of the lead agency and may reject that part of the EIS with which it disagrees, stating publicly when it did not (Section 1506.3(a)).

A cooperating agency with jurisdiction by law (e.g., an agency with independent legal responsibilities with respect to the proposal) has an independent legal obligation to comply with NEPA. Therefore, if the cooperating agency determines that the EIS is wrong or inadequate, it must prepare a supplement to the EIS, replacing or adding any needed information, and must circulate the supplement as a draft for public and agency review and comment. A final supplemental EIS would be required before the agency could take action. The adopted portions of the lead agency EIS should be circulated with the supplement (Section 1506.3(b)). A co-operating agency with jurisdiction by law will have to prepare its own record of decision for its action, in which it must explain how it reached its conclusions. Each agency should explain how and why its conclusions differ, if that is the case, from those of other agencies which issued their records of decision earlier.

An agency that did not cooperate in preparation of an EIS may also adopt an EIS or portion thereof. But this would arise only in rare instances, because an agency adopting an EIS for use in its own decision normally would have been a cooperating agency. If the proposed action for which the EIS was prepared is substantially the same as the proposed action of the adopting agency, the EIS may be adopted as long as it is recirculated as a final EIS and the agency announces what it is doing. This would be followed by the 30-day review period and issuance of a record of decision by the adopting agency. If the proposed action by the adopting agency is not substantially the same as that in the EIS (e.g., if an EIS on one action is being adopted for use in a decision on another action), the EIS would be treated as a draft and circulated for the normal public comment period and other procedures (Section 1506.3(b)).

31a. **Q.** Do the CEQ NEPA regulations apply to independent regulatory agencies like the Federal Energy Regulatory Commission (FERC) and the Nuclear Regulatory Commission (NRC)?

A. The statutory requirements of NEPA Section 102 apply to "all agencies of the federal government." The NEPA regulations implement the procedural provisions of NEPA as set forth in NEPA Section 102(2) for all agencies of the federal government. The NEPA regulations apply to independent regulatory agencies; however, they do not direct independent regulatory agencies or other agencies to make decisions in any particular way or in a way inconsistent with the statutory charter of an agency (Sections 1500.3, 1500.6, 1507.1, and 1507.3).

31b. **Q.** Can an executive branch agency, like the Department of Interior, adopt an EIS prepared by an independent regulatory agency such as FERC?

A. If an independent regulatory agency such as FERC has prepared an EIS in connection with its approval of a proposed project, an executive branch agency (e.g., the Bureau of Land Management in the Department of Interior) may, in accordance with Section 1506.3, adopt the EIS or a portion thereof for its use in considering the same proposal. In such a case, the EIS must, to the satisfaction of the adopting agency, meet the standards for an adequate statement under the NEPA regulations (including scope and quality of analysis of alternatives) and must satisfy the comments and suggestions of the adopting agency. If the independent regulatory agency fails to comply with the NEPA regulations, the cooperating or adopting agency may find that it is unable to adopt the EIS, thus forcing the preparation of a new EIS or EIS supplement for the same action. The NEPA regulations were made applicable to all federal agencies in order to avoid this result and to achieve uniform application and efficiency of the NEPA process.

32. **Q.** Under what circumstances do old EISs have to be supplemented before taking action on a proposal?

A. As a rule of thumb, if the proposal has not yet been implemented, or if the EIS concerns an ongoing program, EISs that are more than 5 years old should be carefully reexamined to determine if the criteria in Section 1502.9 compel preparation of an EIS supplement.

If an agency has made a substantial change in a proposed action that is relevant to environmental concerns, or if there are significant new circumstances or information relevant to environmental concerns and bearing on the proposed action or its impacts, a supplemental EIS must be prepared for an old EIS so that the agency has the best possible information to make any necessary substantive changes in its decisions regarding the proposal. (Section 1502.9(c)).

33a. **Q.** When must a referral of an interagency disagreement be made to the Council?

A. The referral procedure of the Council is a predecision referral process for interagency disagreements. Hence, Section 1504.3 requires that a referring agency must deliver its referral to the Council not later than 25 days after publication by EPA of notice that the final EIS is available (unless the lead agency grants an extension of time under Section 1504.3(b)).

33b. **Q.** May a referral be made after this issuance of a record of decision?

A. No, except for cases where agencies provide an internal appeal procedure which permits simultaneous filing of the final EIS and the record of decision (Section 1506.10(b)(2)). Otherwise, as stated above, the process is a predecision referral process. Referrals must be made within 25 days after the notice of availability of the final EIS, whereas the final decision (ROD) may not be made or filed until after 30 days from the notice of availability of the EIS (Sections 1504.3(b), 1506.10(b)). If a lead agency has granted an extension of time for another agency to take action on a referral, the ROD may not be issued until the extension has expired.

34a. **Q.** Must records of decision be made public? How should they be made available?

A. Under the regulations, agencies must prepare a "concise public record of decision," which contains the elements specified in Section 1505.2. This public record may be integrated into any other decision record prepared by the agency, or it may be separate if decision documents are not normally made public. The record of decision is intended by the Council to be an environmental document (even though it is not explicitly mentioned in the definition of "environmental document" in Section 1508.10). Therefore, it must be made available to the public through appropriate public notice, as required by Section 1506.6(b). However, there is no specific requirements for publication of the ROD itself, either in the *Federal Register* or elsewhere.

34b. **Q.** May the summary section in the final environmental impact statement substitute for or constitute the record of decision of an agency?

A. No. An environmental impact statement is supposed to inform the decision maker **before** the decision is made (Section 1502.1, 1505.2). The CEQ regulations provide for a 30-day period after notice is published that the final EIS has been filed with EPA before the agency may take final action. During that period, in addition to the internal final review of the agency, the public and other agencies can comment on the final EIS prior to the final action of the agency on the proposal. In addition, the CEQ regulations make it clear that the requirements for the summary in an EIS are not the same as the requirements for a ROD (Sections 1502.12 and 1505.2).

34c. **Q.** What provisions should records of decision contain pertaining to mitigation and monitoring?

A. Lead agencies "shall include appropriate conditions (including mitigation measures and monitoring and enforcement programs) in grants, permits or other approvals" and shall "condition funding of actions on mitigation" (Section 1505.3). Any such measures that are adopted must be explained and committed in the ROD.

The reasonable alternative mitigation measures and monitoring programs should have been addressed in the draft and final EIS. The discussion of mitigation and monitoring in a record of decision must be more detailed than a general statement that mitigation is being required, but not so detailed as to duplicate discussion of mitigation in the EIS. The record of decision should contain a concise summary identification of the mitigation measures which the agency has committed itself to adopt.

The record of decision must also state whether all practicable mitigation measures have been adopted and, if not, why not (Section 1505.2(c)). The record of decision must identify the mitigation measures and monitoring and enforcement programs that have been selected and plainly indicate that they are adopted as part of the decision of the agency. If the proposed action is the issuance of a permit or other approval, the specific details of the mitigation measures shall then be included as appropriate conditions in whatever grants, permits, funding or other approvals are being made by the federal agency (Section 1505.3(a),(b)). If the proposal is to be carried out by the federal agency itself, the record of decision should delineate the mitigation and monitoring measures in sufficient detail to constitute an enforceable commitment or incorporate by reference the portions of the EIS that do so.

34d. **Q.** What is the enforceability of a record of decision?

A. Pursuant to generally recognized principles of federal administrative law, agencies will be held accountable for preparing records of decision that conform to the decisions actually made and for carrying out the actions set forth in the records of decision. This is based on the principle that an agency must comply with its own decisions and regulations once they are adopted. Thus, the terms of a record of decision are enforceable by agencies and private parties. A record of decision can be used to compel compliance with or execution of the mitigation measures identified therein.

35. **Q.** How long should the NEPA process take to complete?

A. When an EIS is required, the process obviously will take longer than when an EA is the only document prepared. But the CEQ NEPA regulations encourage streamlined review, adoption of deadlines, elimination of duplicative work, eliciting suggested alternatives and other comments early through scoping, cooperation among agencies, and consultation with applicants during project planning. The Council has advised agencies that under the new NEPA regulations, even large, complex energy projects would require only about 12 months for the completion of the entire EIS process. For most major actions, this period is well within the planning time that is needed in any event, apart from NEPA.

The time required for the preparation of program EISs may be greater. The Council also recognizes that some projects will entail difficult long-term planning and/or the acquisition of certain data which of necessity will require more time for the preparation of the EIS. Indeed, some proposals should be given more time for the thoughtful preparation of an EIS and development of a decision which fulfills the substantive goals of NEPA.

For cases in which only an environmental assessment will be prepared, the NEPA process should take no more than 3 months, and in many cases, substantially less, as part of the normal analysis and approval process for the action.

36a. **Q.** How long and detailed must an environmental assessment (EA) be?

A. The environmental assessment is a concise public document which has three defined functions: (1) it briefly provides sufficient evidence and analysis for determining whether to prepare an EIS; (2) it aids the compliance of an agency with NEPA when no EIS is necessary, i.e., it helps to identify better alternatives and mitigation measures; and (3) it facilitates preparation of an EIS when one is necessary (Section 1508.9(a)).

Since the EA is a concise document, it should not contain long descriptions or detailed data which the agency may have gathered. Rather, it should contain a brief discussion of the need for the proposal, alternatives to the proposal, the environmental impacts of the proposed action and alternatives, and a list of agencies and persons consulted (Section 1508.9(b)).

While the regulations do not contain page limits for EAs, the Council has generally advised agencies to keep the length of EAs to not more than approximately 10 to 15 pages. Some agencies expressly provide page guidelines (e.g., 10 to 15 pages in the case of the Army Corps). To avoid undue length, the EA may incorporate by reference background data to support its concise discussion of the proposal and relevant issues.

36b. **Q.** Under what circumstances is a lengthy EA appropriate?

A. Agencies should avoid preparing lengthy EAs, except in unusual cases where a proposal is so complex that a concise document cannot meet the goals of Section 1508.9 and where it is extremely difficult to determine whether the proposal could have significant environmental effects. In most cases, however, a lengthy EA indicates that an EIS is needed.

37a. **Q.** What is the level of detail of information that must be included in a finding of no significant impact (FONSI)?

A. The FONSI is a document in which the agency briefly explains the reasons why an action will not have a significant effect on the human environment and, therefore, why an EIS will not be prepared (Section 1508.13). The finding itself need not be detailed, but must succinctly state the reasons for deciding that the action will have no significant environmental effects, and, if relevant, must show which factors were weighted most heavily in the determination. In addition to this statement, the FONSI must include, summarize, or attach and incorporate by reference the environmental assessment.

37b. **Q.** What are the criteria for deciding whether a FONSI should be made available for public review for 30 days before the final determination of the agency or whether to prepare an EIS?

A. Public review is necessary, for example, (1) if the proposal is a borderline case, i.e., when there is a reasonable argument for preparation of an EIS; (2) if it is an unusual case, a new kind of action, or a precedent-setting case such as a first intrusion of even a minor development into a pristine area; (3) when there is either scientific or public controversy over the proposal; or (4) when it involves a proposal which is or is closely similar to one which normally requires preparation of an EIS (Sections 1501.4(c)(2), 1508.27). Agencies also must allow a period of public review of the FONSI if the proposed action would be located in a floodplain or wetland (E.O. 11988, Sec. 2(a)(4); E.O. 11990, Sec. 2(b)).

38. **Q.** Must EAs and FONSIs be made public? If so, how should this be done?

A. Yes, they must be available to the public. Section 1506.6 requires agencies to involve the public in implementing their NEPA procedures, and this includes public involvement in the preparation of EAs and FONSIs. These are public ''environmental documents'' under Section 1506.6(b), and, therefore, agencies must give public notice of their availability. A combination of methods may be used to give notice, and the methods should be tailored to the needs of particular cases. Thus, a *Federal Register* notice of availability of the documents, coupled with notices in national publications and mailed to interested national groups might be appropriate for proposals that are national in scope. Local newspaper notices may be more appropriate for regional or site-specific proposals.

The objective, however, is to notify all interested or affected parties. If this is not being achieved, then the methods should be reevaluated and changed. Repeated failure to reach the interested or affected public would be interpreted as a violation of the regulations.

39. **Q.** Can an EA and FONSI be used to improve enforceable mitigation measures, monitoring programs, or other requirements, even though there is no requirement in the regulations in such cases for a formal record of decision?

A. Yes. In cases where an environmental assessment is the appropriate environmental document, there still may be mitigation measures or alternatives that would be desirable to consider and adopt, even though the impacts of the proposal will not be "significant". In such cases, the EA should include a discussion of these measures or alternatives to "assist agency planning and decision making" and to "aid an agency's compliance with (NEPA) when no environmental impact statement is necessary" (Section 1501.3(b), 1508.9(2)). The appropriate mitigation measures can be imposed as enforceable permit conditions, or adopted as part of the final decision of the agency in the same manner mitigation measures are adopted in the formal record of decision that is required in EIS cases.

40. **Q.** If an environmental assessment indicates that the environmental effects of a proposal are significant, but that, with mitigation, those effects may be reduced to less-than-significant levels, may the agency make a finding of no significant impact rather than prepare an EIS. Is that a legitimate function of an EA and scoping?

A. Mitigation measures may be relied upon to make a finding of no significant impact only if they are imposed by statute or regulation or submitted by an applicant or agency as part of the original proposal. As a general rule, the regulations contemplate that agencies should use a broad approach in defining significance, and should not rely on the possiblity of mitigation as an excuse to avoid the EIS requirement (Sections 1508.8, 1508.27).

If a proposal appears to have adverse effects which would be significant, and certain mitigation measures are then developed during the scoping or EA stages, the existence of such possible mitigation does not obviate the need for an EIS. Therefore, if scoping or the EA identifies certain mitigation possibilities without altering the nature of the overall proposal itself, the agency should continue the EIS process and submit the proposal and the potential mitigation for public and agency reviews and comment. This is essential to ensure that the final decision is based on all the relevant factors, and that the full NEPA process will result in enforceable mitigation measures through the record of decision.

In some instances where the proposal itself so integrates mitigation from the beginning that it is impossible to define the proposal without including the mitigation, the agency may then rely on the mitigation measures in determining that the overall effects would not be significant (e.g., where an application for a permit for a small hydroelectric dam is based on a binding commitment to build fish ladders, to permit adequate down stream flow, and to replace any lost wetlands, wildlife habitat, and recreational potential). In those instances, agencies should make the FONSI and EA available for 30 days of public comment before taking action (Section 1501.4(e)(2)).

Similarly, scoping may result in a redefinition of the entire project as a result of mitigation proposals. In that case, the agency may alter its previous decision to do an EIS, as long as the agency or applicant resubmits the entire proposal and the EA and FONSI are available for 30 days of review and comment. One example of this would be where the size and location of a proposed industrial park are changed to avoid affecting a nearby wetland area.

Chapter 11

CONTROL TECHNOLOGY BIBLIOGRAPHY

A very large amount of useful data and information on control technology has been published in the form of books, individual scientific and technical articles, and major conference proceedings since 1970. Examples of the control technology literature published in 1985 and 1986 are provided below. These papers are extracted with the extensive ''Technical Papers Reference File'' compiled by the editors of *Power* magazine and published in the November 1986 issue of *Power,* a monthly publication of McGraw-Hill, Inc., New York, N.Y. 10020. The following list is only an example of the broad base of data in pollution control technology applicable to coal combustion, oil combustion, and hazardous waste.

1. ''**Catalytic thermal destruction**'', H. Green, University of Akron, in the 1986 Hazardous Waste Research Symposium Proceedings, U.S. Environmental Protection Agency, Washington, D.C.
2. ''**Practical limitations of waste characteristics for effective incineration**'', E. Martin, in the 1986 Hazardous Waste Research Symposium Proceedings, U.S. Environmental Protection Agency, Washington, D.C.
3. ''**Incinerator and cement kiln capacity for hazardous waste treatment**'', G. Vogel, Mitre Corp., in the 1986 Hazardous Waste Research Symposium Proceedings, U.S. Environmental Protection Agency, Washington, D.C.
4. ''**Non-steady state testing of industrial boilers burning hazardous wastes**'', C. Castaldini, Acurex Corp., in the 1986 Hazardous Waste Research Symposium Proceedings, U.S. Environmental Protection Agency, Washington, D.C.
5. ''**Experience from retrofitting a dry scrubbing system at an existing industrial powerplant**'' (ASME Book No. I00195), N. F. Tanagho, Rockwell International, et al., in the 1986 Industrial Power Conference Proceedings, American Society of Mechanical Engineers, 22 Law Drive, Fairfield, N.J. 07007.
6. ''**Combustion upgrade and particulate removal on a bark-fired power boiler**'', (ASME Book No. I00195), B. W. Seidel, BE & K Engineering Co., et al., in the 1986 Industrial Power Conference Proceedings, American Society of Mechanical Engineers, 22 Law Drive, Fairfield, N.J. 07007.
7. ''**Reduction of SO_2 emissions from coal-fired facilities**'', H. Feibus, in the 1986 Environmental Technology Conference, 966 Hungerford Drive, Rockville, Md. 20850.
8. ''**Status of retrofit SO_2 control technologies**'', J. Maulbetsch, in the 1986 Environmental Technology Conference, 966 Hungerford Drive, Rockville, Md. 20850.
9. ''**Control of air emissions from resource recovery facilities**'', J. Hahn, Cooper Engineers, in the 1986 Environmental Technology Conference, 966 Hungerford Drive, Rockville, Md. 20850.
10. ''**Performance and economics of a spray dryer FGD system used with high-sulfur coal**'', C. D. Livengood, Argonne National Laboratory, et al., in the 1986 American Power Conference Proceedings, Illinois Institute of Technology, Chicago, Ill. 60616.
11. ''**Evaluation of retrofit options for SO_2 and NO_x control for an Eastern electric utility**'', W. F. Frazier, Virginia Electric & Power Co., in the 1986 American Power Conference Proceedings, Illinois Institute of Technology, Chicago, Ill. 60616.
12. ''**Spray dryer/ESP testing for utility retrofit applications on high-sulfur coal**'', R. F. Robards, TVA, et al., in the 1986 American Power Conference Proceedings, Illinois Institute of Technology, Chicago, Ill. 60616.
13. ''**Precipitator or baghouse? Evaluation to retrofit for no visible emissions on low-**

sulfur coal'', R. R. Gosnell, III, Bechtel Power Corp., et al., in the 1986 American Power Conference Proceedings, Illinois Institute of Technology, Chicago, Ill. 60616.

14. ''**Current status of in-duct dry flue-gas-desulfurization: A cost effective retrofit technology for reduction of SO emissions**'', N. Z. Shilling, General Electric Environmental Services, Inc., in the American Power Conference Proceedings, Illinois Institute of Techology, Chicago, Ill. 60616.
15. ''**Acid dew point: New concern for an old problem in fossil plant design**'', R. H. Sirois, Stone & Webster Engineering Corp., et al., in the 1986 American Power Conference Proceedings, Illinois Institute of Technology, Chicago, Ill. 60616.
16. ''**Selective catalytic reduction — impact on heat-recovery steam generator design and operation**'', (ASME Paper No. 86-GR-72), J. S. Davis, Struthers Wells Corp., et al., in the 1986 Gas Turbine Conference Proceedings, American Society of Mechanical Engineers, 22 Law Drive, Fairfield, N.J. 07007.
17. ''**Methods and systems for applying sonic power to baghouse cleaning**'', D. Pontius, et al., in the 1985 Fabric Filter Conference Proceedings, Electric Power Research Institute (EPRI), Palo Alto, Calif. 94303.
18. ''**Flyash penetration in periodically cleaned woven fabric filters**'', W. Davis, University of Tennessee, et al., in the 1985 Fabric Filter Conference Proceedings, Electric Power Research Institute (EPRI), Palo Alto, Calif. 94303.
19. ''**Relationship between fabric properties and baghouse performance**'', W. Smith, et al., in the 1985 Fabric Filter Conference Proceedings, Electric Power Research Institute (EPRI), Palo Alto, Calif. 94303.
20. ''**Operating experience of a fabric filter on a cyclone boiler**'', N. Shilling, General Electric Environmental Services Inc., et al., in the 1985 Fabric Filter Conference Proceedings, Electric Power Research Institute (EPRI), Palo Alto, Calif. 94303.
21. ''**Fabric filter operating experience at Arizona Public Service Co.'s Four Corners Units 4 an 5**'', U. Singh, Arizona Public Service Co., et al., in the 1985 Fabric Filter Conference Proceedings, Electric Power Research Institute (EPRI), Palo Alto, Calif. 94303.
22. ''**Experience with pulse-jet-cleaned fabric filters on Australian coal-fired utility boilers**'', C. Barranger, Flakt U.S.A., et al., in the 1985 Fabric Filter Conference Proceedings, Electric Power Research Institute (EPRI), Palo Alto, Califo. 94303.
23. ''**Improving baghouse performance at the Monticello station**'', L. Felix, Southwest Research Institute, et al., in the 1985 Fabric Filter Conference Proceedings, Electric Power Research Institute (EPRI), Palo Alto, Calif. 94303.
24. ''**Sonic horn testing at the Tolk station**'', J. Hudspeth, Southwestern Public Service Co., in the 1985 Fabric Filter Conference Proceedings, Electric Power Research Institute (EPRI), Palo Alto, Calif. 94303.
25. ''**Dry sorbent injection for SO_2 removal**'', R. Rhudy, EPRI, et al., in the 1985 Fabric Filter Conference Proceedigns, Electric Power Research Institute (EPRI), Palo Alto, Calif. 94303.
26. ''**Effect of particulate control device selection on SO_2 scrubber performance**'', P. Maroney, Brown & Caldwell, et al., in the 1985 Fabric Filter Conference Proceedings, Electric Power Research Institute (EPRI), Palo Alto, Calif. 94303.
27. ''**Utilization of coal gasification for power generation**'', T. J. Marqueen, Ebasco Services Inc., in the 1985 Coal Technology Conference Proceedings, P.O. Box 721948, Houston, Tex. 77272.
28. ''**SO_2-removal system improvement program at seven units of Louisville Gas & Electric Co.**'', D. A. Froelich, Burns & McDonnell Engineering Co., et al., in the 1985 Coal Technology Conference Proceedings, P.O. Box 721948, Houston, Tex. 77272.

29. "**Dust collection/suppression in the Belle River plant coal-handling system**", J. A. Bonner, Bechtel Associates Professional Corp., et al., in the 1985 Coal Technology Conference Proceedings, P.O. Box 721948, Houston, Tex. 77272.
30. "**Pulse energization test results for 125- and 600-MW utility electrostatic precipitators**", A. A. Schuetz, Combustion Engineering Inc., et al., in the 1985 Coal Technology Conference Proceedings, P.O. Box 721948, Houston, Tex. 77272.
31. "**FGD byproduct gypsum — Board of Light & Power, Grand Haven, Mich.**", D. W. Wilterdink, Grand Haven Board of Light & Power, et al., in the 1985 Coal Technology Conference Proceedings, P.O. Box 721948, Houston, Tex. 77272.
32. "**Optimized lime or other additives in FGD sludge stabilization**", J. C. Parks, Burns & McDonnell Engineering Co., et al., in the 1985 Coal Technology Conference Proceedings, P.O. Box 721948, Houston, Tex. 77272.
33. "**Feasibility of converting an FGD system from lime to limestone as a reagent**", J. Landwehr, Burns & McDonnell Engineering Co., et al., in the 1985 Coal Technology Conference Proceedings, P.O. Box 721948, Houston, Tex. 77272.
34. "**Using synthetic liners to control runoffs from coal piles**", B. Ainsworth, Schlegel Lining Technology, in the 1985 Coal Technology Conference Proceedings, P.O. Box 721948, Houston, Tex. 77272.
35. "**Design and development of a retrofit low NO_x burner**", I. J. Stuart-Sheppard, Ontario Hydro, et al., in the 1985 Coal Technology Conference Proceedings, P.O. Box 721948, Houston, Tex. 77272.
36. "**When hazardous waste threatens the groundwater**", C. R. Ryan, Geo Con Inc., in the 1985 International Water Conference Proceedings, Engineer's Society of Western Pennsylvania, Pittsburgh, Pa. 15219.
37. "**Vanadium removal from oil-fired powerplant wastewater**", R. Rosain, CH_2M Hill Inc., et al, in the 1985 International Water Conference Proceedings, Engineer's Society of Western Pennsylvania, Pittsburgh, Pa. 15219.
38. "**Water-related problems in ash transport systems**", P. A. Barker, Nalco Chemical Co., et al., in the 1985 International Water Conference Proceedings, Engineer's Society of Western Pennsylvania, Pittsburgh, Pa. 15219.
39. "**Radwaste ion-exchange optimization**", N. P. Jacob, Babcock & Wilcox Co., et al., in the 1985 International Water Conference Proceedings, Engineer's Society of Western Pennsylvania, Pittsburgh, Pa. 15219.
40. "**Zero blowdown by sidestream softening on cooling systems of base load units**", R. Proctor, Sierra Pacific Power Co., in the 1985 International Water Conference Proceedings, Engineer's Society of Western Pennsylvania, Pittsburgh, Pa. 15219.
41. "**Onstream iron removal**", G. D. Engstrom, Dearborn Chemical Co., et al., in the 1985 International Water Conference Proceedings, Engineer's Society of Western Pennsylvania, Pittsburgh, Pa. 15219.
42. "**Initial operating experience with a liquid waste-treatment system for NYSE&G's Somerset Unit 1**", R. S. Bernstein, Ebasco Services Inc., et al., in the 1985 International Water Conference Proceedings, Engineer's Society of Western Pennsylvania, Pittsburgh, Pa. 15219.
43. "**Combustion generated particulate emissions**" (ASME Book No. I00210), K. L. Tuttle, U.S. Naval Academy, in the 1985 National Waste Processing Conference Proceedings, American Society of Mechanical Engineers, 22 Law Drive, Fairfield, N.J. 07007.
44. "**Innovative technology for the control of air pollution at waste-to-energy plants**" (ASME Book No. I00210), J. L. Hahn, Cooper Engineers Inc., in the 1985 National Waste Processing Conference Proceedings, American Society of Mechanical Engineers, 22 Law Drive, Fairfield, N.J. 07007.

45. ‘‘**Minimizing trace organic emissions from combustion of municipal wastes by the use of carbon monoxide monitors**’’ (ASME Book No. I00210), F. Hasselriis, Gershman, Brickner & Bratton, in the 1985 National Waste Processing Conference Proceedings, American Society of Mechanical Engineers, 22 Law Drive, Fairfield, N.J. 07007.
46. ‘‘**Energy saving in sewage sludge incineration with indirect heat dryer**’’ (ASME Book No. I00210), K. Okazawa, Mitsubishi Heavy Industries Ltd., et al., in the 1985 National Waste Processing Conference Proceedings, American Society of Mechanical Engineers, 22 Law Drive, Fairfield, N.J. 07007.
47. ‘‘**Incineration of hot sludge**’’ (ASME Book No. I00210), C. R. Brunner, CH_2M Hill Inc., et al., in the 1985 National Waste Processing Conference Proceedings, American Society of Mechanical Engineers, 22 Law Drive, Fairfield, N.J. 07007.
48. ‘‘**Operating experience and performance of wet ash-handling systems used in mass-burning solid-waste facilities**’’ (ASME Book No. I00210), J. Stelian, Beaumont Birch Co., et al., in the 1985 National Waste Processing Conference Proceedings, American Society of Mechanical Engineers, 22 Law Drive, Fairfield, N.J. 07007.
49. ‘‘**Design considerations for FGD dampers**’’ (ASME Paper No. 85-JPGC-PWR-47), A. W. Ferguson, Black & Veatch, et al., in the 1985 Joint Power Generation Conference, American Society of Mechanical Engineers, 22 Law Drive, Fairfield, N.J. 07007.
50. ‘‘**Carbon particulate reduction evaluation for oil-fired utility boilers**’’ (ASME Paper No. 85-JPGC-APC-4), J. Nylander, San Diego Gas & Electric et al., in the 1985 Joint Power Generation Conference, American Society of Mechanical Engineers, 22 Law Drive, Fairfield, N.J. 07007.
51. ‘‘**Economical, innovative methods of straightening ESP collector plates**’’ (ASME Paper No. 85-JPGC-APC-5), R. E. Jonelis, Midwest Power Corp., in the 1985 Joint Power Generation Conference, American Society of Mechanical Engineers, 22 Law Drive, Fairfield, N.J. 07007.
52. ‘‘**Applications of spray dryer/baghouse technology to emissions from incinerators**’’ (ASME Paper No. 85-JPGC-APC-14), D. C. Borio, Combustion Engineering Inc., et al., in the 1985 Joint Power Generation Conference, American Society of Mechanical Engineers, 22 Law Drive, Fairfield, N.J. 07007.
53. ‘‘**Particulate control technology technical committee annual report 1985**’’ (ASME Paper No. 85-JPGC-APC-10), T. Bechtel, General Electric Environmental Services., in the 1985 Joint Power Generation Conference, American Society of Mechanical Engineers, 22 Law Drive, Fairfield, N.J. 07007.
54. ‘‘**Fluidized-bed-combustion: A proven alternative**’’ (ASME Paper No. 85-JPGC-APC-11), N. B. Smith, Stanley Consultants Inc., et al., in the 1985 Joint Power Generation Conference, American Society of Mechanical Engineers, 22 Law Drive, Fairfield, N.J. 07007.
55. ‘‘**Fuel oil reburning application for NO_x control to firetube package boilers**’’ (ASME Paper No. 85-JPGC-APC-6), J. A. Mulholland, U.S. EPA, et al., in the 1985 Joint Power Generation Conference, American Society of Mechanical Engineers, 22 Law Drive, Fairfield, N.J. 07007.
56. ‘‘**Solid and liquid waste technical committee annual report 1985**’’ (ASME Paper No. 85-JPGC-APC-9), W. Ellison, Ellison Consultants, in the 1985 Joint Power Generation Conference, American Society of Mechanical Engineers, 22 Law Drive, Fairfield, N.J. 07007.
57. ‘‘**Air quality and resource recovery: The major regulatory issues**’’, C. Kemp, Canruf Co., in the 1985 Conference Energy from Municipal Waste, October 23, 1985, Proceedings, *Power* magazine, New York, N.Y. 10020.

58. ''**Minimizing refuse combustion emissions by combustion control, alkaline reagents, condensation, and particulate removal**'', F. Hasselriis, Consultant, in the 1985 Conference Energy from Municipal Waste, October 23, 1985, Proceedings, *Power* magazine, New York, N.Y. 10020.
59. ''**Use of dry sorbent injection for SO_2 emission control of coal-fired boilers and the potential reuse of the waste chemicals in the draft recovery cycle**'', T. W. Sonnichsen, Arie Verloop, et al., in the 1985 TAPPI Engineering Conference Proceedings, Volume 1, Technical Association of the Paper and Pulp Industry, Atlanta, Ga. 30348-5113.
60. ''**Storage of chemically stabilized FGD waste**'', L. L. Holish, Paper published by Sargent & Lundy, 55 E. Monroe St., Chicago, Ill. 60603.
61. ''**Treatment and disposal of wastes from dry SO_2 control processes**'', S. M. Katzberger, et al., Paper published by Sargent & Lundy, 55 E. Monroe St., Chicago, Ill. 60603.
62. ''**Particle class analysis for coal-fired powerplant flyash**'' (APCA Paper No. 86-40.5), D. S. Kim, University of Illinois, et al., in the 1986 Air Pollution Control Association Conference Proceedings, Air Pollution Control Association, P.O. Box 2861, Pittsburgh, Pa. 15230 (order by paper number).
63. ''**Emission control technologies for resource recovery**'' (APCA Paper No. 86-50.11), M. J. Clarke, New York City Department of Sanitation, in the 1986 Air Pollution Control Association Conference Proceedings, Air Pollution Control Association, P.O. Box 2861, Pittsburgh, Pa. 15230 (order by paper number).
64. ''**Layout criteria for a refuse incineration plant to minimize the impact on the environment**'' (APCA Paper No. 86-50.7), K. Horch, Deutsche Babcock Anlagen AG, et al., in the 1986 Air Pollution Control Association Conference Proceedings, Air Pollution Control Association, P.O. Box 2861, Pittsburgh, Pa. 15230 (order by paper number).
65. ''**Dioxin emissions from modern, mass-fired, stoker/boilers for use in waste-to-energy risk assessment**'' (APCA Paper No. 86-50.12), J. L. Hahn, Ogden Martin Systems Inc., in the 1986 Air Pollution Control Association Conference Proceedings, Air Pollution Control Association, P.O. Box 2861, Pittsburgh, Pa. 15230 (order by paper number).
66. ''**Incineration: Waste management for the 80s**'' (APCA Paper No. 86-31.2), T. Shen, Columbia University et al., in the 1986 Air Pollution Control Association Conference Proceedings, Air Pollution Control Association, P.O. Box 2861, Pittsburgh, Pa. 15230 (order by paper number).
67. ''**Controlling acid deposition by seasonal gas substitution in coal-fired powerplants**'' (APCA Paper No. 86-31.4), D. S. Colomb, Massachusetts Institute of Technology, et al., in the 1986 Air Pollution Control Association Conference Proceedings, Air Pollution Control Association, P.O. Box 2861, Pittsburgh, Pa. 15230 (order by paper number).
68. ''**Application of the optimized environmental planning program to a hypothetical utility affected by proposed tall stacks regulations**'' (APCA Paper No. 86-56.4), J. A. Brander, General Electric Co., in the 1986 Air Pollution Control Association Conference Proceedings, Air Pollution Control Association, P.O. Box 2861, Pittsburgh, Pa. 15230 (order by paper number).
69. ''**Nearby terrain effects in a good engineering practice stack height demonstration**'' (APCA Paper No. 86-56.5), S. D. Taylor, Kennecott Corp., Utah Copper Div., et al., in the 1986 Air Pollution Control Association Conference Proceedings, Air Pollution Control Association, P.O. Box 2861, Pittsburgh, Pa. 15230 (order by paper number).

70. ''**Unified wet deposition database for Eastern North America: Data screening and calculation procedures and results**'' (APCA Paper No. 86-29.1), M. Lusis, Ontario Ministry of Environment, in the 1986 Air Pollution Control Association Conference Proceedings, Air Pollution Control Association, P.O. Box 2861, Pittsburgh, Pa. 15230 (order by paper number).
71. ''**Conceptual design and chemical data needs for evaluation of regional acidic deposition models**'' (APCA Paper No. 86-29.6), R. L. Dennis, U.S. EPA, et al., in the 1986 Air Pollution Control Association Conference Proceedings, Air Pollution Control Association, P.O. Box 2861, Pittsburgh, Pa. 15230 (order by paper number).
72. ''**New process for emission control and waste heat recovery from a hazardous waste incinerator**'' (APCA Paper No. 86-61.6), J. D. Brady, Anderson 2000 Inc., in the 1986 Air Pollution Control Association Conference Proceedings, Air Pollution Control Association, P.O. Box 2861, Pittsburgh, Pa. 15230 (order by paper number).
73. ''**Application of dry flue gas scrubbing to hazardous waste incineration**'' (APCA Paper No. 86-10.4), P. J. Kroll, International Waste Energy System, et al., in the 1986 Air Pollution Control Association Conference Proceedings, Air Pollution Control Association, P.O. Box 2861, Pittsburgh, Pa. 15230 (order by paper number).
74. ''**Control technology for MSW incinerator applications**'' (APCA Paper No. 86-82.1), R. W. McIlvaine, McIlvaine Co., in the 1986 Air Pollution Control Association Conference Proceedings, Air Pollution Control Association, P.O. Box 2861, Pittsburgh, Pa. 15230 (order by paper number).
75. ''**Equipment design considerations for the control of emissions from waste-to-energy facilities**'' (APCA Paper No. 86-82.4), W. B. Ferguson, Combustion Engineering Inc., et al., in the 1986 Air Pollution Control Association Conference Proceedings, Air Pollution Control Association, P.O. Box 2861, Pittsburgh, Pa. 15230 (order by paper number).
76. ''**Design considerations for resource recovery spray dryer absorption systems**'' (APCA Paper No. 86-82.6), J. R. Donnelly, Joy Manufacturing Co., et al., in the 1986 Air Pollution Control Association Conference Proceedings, Air Pollution Control Association, P.O. Box 2861, Pittsburgh, Pa. 15230 (order by paper number).
77. ''**Design of a specific system for waste solvent combustion**'', (APCA Paper No. 86-49.2), A. W. Driessen, Nashua Corp., et al., in the 1986 Air Pollution Control Association Conference Proceedings, Air Pollution Control Association, P.O. Box 2861, Pittsburgh, Pa. 15230 (order by paper number).
78. ''**Acid rain issue in Italy**'' (APCA Paper No. 86-34.2), E. Ioannilli, Centro Ricerche Termiche, et al., in the 1986 Air Pollution Control Association Conference Proceedings, Air Pollution Control Association, P.O. Box 2861, Pittsburgh, Pa. 15230 (order by paper number).
79. ''**Acid rain in Puerto Rico**'' (APCA Paper No. 86-34.5), R. M. Candelario, University of Puerto Rico, in the 1986 Air Pollution Control Association Conference Proceedings, Air Pollution Control Association, P.O. Box 2861, Pittsburgh, Pa. 15230 (order by paper number).
80. ''**Emission assessment for a 169-MW AFB combustor**'' (APCA Paper No. 86-47.2), D. Lokey, TVA, et al., in the 1986 Air Pollution Control Association Conference Proceedings, Air Pollution Control Association, P.O. Box 2861, Pittsburgh, Pa. 15230 (order by paper number).
81. ''**Emissions from fluidized bed combustion**'' (APCA Paper No. 86-49.5), R. W. Shedd, Stone Johnston Corp., in the 1986 Air Pollution Control Association Conference Proceedings, Air Pollution Control Association, P.O. Box 2861, Pittsburgh, Pa. 15230 (order by paper number).
82. ''**Development of an independent 50-state combustion source fuel and emissions**

inventory for analysis of new energy technologies and acid deposition'' (APCA Paper No. 86-30.5), M. F. Szabo, PEI Associates Inc., in the 1986 Air Pollution Control Association Conference Proceedings, Air Pollution Control Association, P.O. Box 2861, Pittsburgh, Pa. 15230 (order by paper number).

83. ``**Meeting atmospheric modeling requirements for utility data**'' (APCA Paper No. 86-30.8), E. H. Pechan, E. H. Pechan & Associates, Inc., et al., in the 1986 Air Pollution Control Association Conference Proceedings, Air Pollution Control Association, P.O. Box 2861, Pittsburgh, Pa. 15230 (order by paper number).

84. ``**Simultaneous control of SO_2 and NO_x in spray dry scrubbing**'' (APCA Paper No. 86-47.4), R. Chan, University of Illinois, et al., in the 1986 Air Pollution Control Association Conference Proceedings, Air Pollution Control Association, P.O. Box 2861, Pittsburgh, Pa. 15230 (order by paper number).

85. ``**Low capital cost, retrofit SO_2 control technologies for high-sulfur coal applications**'' (APCA Paper No. 86-47.5), H. Yoon, Conoco Coal Research Div., et al., in the 1986 Air Pollution Control Association Conference Proceedings, Air Pollution Control Association, P.O. Box 2861, Pittsburgh, Pa. 15230 (order by paper number).

86. ``**Statistical design for screening in NO_x/SO_x environmental control technology**'' (APCA Paper No. 86-47.6), J. B. L. Harkness, Argonne National Laboratory, in the 1986 Air Pollution Control Association Conference Proceedings, Air Pollution Control Association, P.O. Box 2861, Pittsburgh, Pa. 15230 (order by paper number).

87. ``**Technical and economic aspects of a pressurized circulating FBC system**'' (APCA Paper No. 86-41.1), S. R. Wysk, Combustion Engineering Inc., et al., in the 1986 Air Pollution Control Association Conference Proceedings, Air Pollution Control Association, P.O. Box 2861, Pittsburgh, Pa. 15230 (order by paper number).

88. ``**Status of the DOE/General Electric Environmental Services Inc. in-duct scrubbing pilot study**'' (APCA Paper No. 86-41.4), K. R. Murphy, et al., General Electric Environmental Services, in the 1986 Air Pollution Control Association Conference Proceedings, Air Pollution Control Association, P.O. Box 2861, Pittsburgh, Pa. 15230 (order by paper number).

89. ``**Application of reburning for NO_x control to coal- and oil-fired furnaces**'' (APCA Paper No. 86-73.1), J. M. McCarthy, Energy & Environmental Research Corp., et al., in the 1986 Air Pollution Control Association Conference Proceedings, Air Pollution Control Association, P.O. Box 2861, Pittsburgh, Pa. 15230 (order by paper number).

90. ``**Fluidized slurry wet scrubbing technique**'' (APCA Paper No. 86-73.4), K. C. Schifftner, Chem-Pro Corp., in the 1986 Air Pollution Control Association Conference Proceedings, Air Pollution Control Association, P.O. Box 2861, Pittsburgh, Pa. 15230 (order by paper number).

91. ``**Joy/Niro control technology for hazardous waste incinerators**'' (APCA Paper No. 86-87.6), J. R. Donnelly, Joy Manufacturing Co., et al., in the 1986 Air Pollution Control Association Conference Proceedings, Air Pollution Control Association, P.O. Box 2861, Pittsburgh, Pa. 15230 (order by paper number).

92. ``**Treatment and disposal of hazardous wastes in the Federal Republic of Germany**'' (APCA Paper No. 86-87.8), P. Schoner, Hessiche Industriemull GmbH, in the 1986 Air Pollution Control Association Conference Proceedings, Air Pollution Control Association, P.O. Box 2861, Pittsburgh, Pa. 15230 (order by paper number).

93. ``**Determination of long range transport potential of NO_x and SO_x on the northeastern U.S.**'' (APCA Paper No. 86-48.5), S. T. Thomas, University of Toledo, et al., in the 1986 Air Pollution Control Association Conference Proceedings, Air Pollution Control Association, P.O. Box 2861, Pittsburgh, Pa. 15230 (order by paper number).

94. ''**Source apportionment of eastern U.S.-Canada transboundary flows of sulfur and nitrogen**'' (APCA Paper No. 86-48.8), J. A. Fay, Massachusetts Institute of Technology, et al., in the 1986 Air Pollution Control Association Conference Proceedings, Air Pollution Control Association, P.O. Box 2861, Pittsburgh, Pa. 15230 (order by paper number).
95. ''**Regional scale transport of acid rain precursors from a coal-fired powerplant: Enhanced transport by a nocturnal jet**'' (APCA Paper No. 86-48.10), D. D. Reible, Louisiana State University, et al., in the 1986 Air Pollution Control Association Conference Proceedings, Air Pollution Control Association, P.O. Box 2861, Pittsburgh, Pa. 15230 (order by paper number).
96. ''**FGD operation and maintenance cost: A function of first cost — Or is it?**'', (APCA Paper No. 86-63.4), D. C. Gaige, Burns & McDonnell Engineering Corp., et al., in the 1986 Air Pollution Control Association Conference Proceedings, Air Pollution Control Association, P.O. Box 2861, Pittsburgh, Pa. 15230 (order by paper number).
97. ''**Brandon Woods Energy Business Park structural flyash fill**'' (APCA Paper No. 86-88.6), D. W. Parks, Baltimore Gas & Electric Co., et al., in the 1986 Air Pollution Control Association Conference Proceedings, Air Pollution Control Association, P.O. Box 2861, Pittsburgh, Pa. 15230 (order by paper number).
98. ''**Characteristics affecting design considerations for waste management systems for advanced SO_2 control technologies**'' (APCA Paper No. 86-88.7), A. G. Eklund, Radian Corp., et al., in the 1986 Air Pollution Control Association Conference Proceedings, Air Pollution Control Association, P.O. Box 2861, Pittsburgh, Pa. 15230 (order by paper number).

REFERENCES

1. **Allen, T. B.,** *Vanishing Wildlife of North America,* (Special Publ. Ser.), National Geographic Society, Washington, D.C., 1974.
2. **Benefield, L. D., Judkins, J. F., and Weand, B. L.,** *Process Chemistry for Water and Wastewater Treatment,* Prentice-Hall, Inc., Englewood Cliffs, N.J., 1982.
3. **Bovine, J. E. and McGarity, T. O.,** *The Law of Environmental Protection: Cases, Legislation, Policies,* West, St. Paul, 1984.
4. Environmental Impact Statements: An Analysis of Six Years' Experience by Seventy Federal Agencies, Council on Environmental Quality, Washington, D.C., 1976.
5. Environmental Quality 1980, 11th Annu. Rep., Council on Environmental Quality, Washington, D.C., 1980.
6. Environmental Quality, 15th Annu. Rep. of the Council on Environmental Quality, Council on Environmental Quality, Washington, D.C., 1984.
7. **Covert, B. C.,** The other type of technical editor, in *Proc. 32nd Int. Technical Communications Conf.,* Barnow, R., Ed., Society for Technical Communications, Washington, D.C., 1985.
8. **Culp, R. L. and Wesner, G. M.,** *Handbook of Advanced Wastewater Treatment,* 2nd ed., Van Nostrand Reinhold, New York, 1978.
9. **Dickert, T. G. and Tuttle, A. E.,** Cumulative impact assessment in environmental planning: a coastal wetland watershed example, *Environ. Impact Assess. Rev.,* 5, 1, 1985.
10. **Eckenfelder, W. W., Jr. and Santhanam, C. I., Eds.,** *Sludge Treatment,* Marcel Dekker, New York, 1981.
11. **Engdahl, R. B. and Murthy, K. S.,** Updating the Applicability of Scrubbers, talk delivered at the Fourth Int. Pollution Control Cong., Cleveland, Ohio, October 20 to 23, 1975.
12. Protection and Enhancement of Environmental Quality, Executive Order 11514, 1970, as amended by Executive Order 11991, 1977.
13. Final Environmental Impact Statement: Cross Steam Plant at Santee-Cooper Project No. 199, South Carolina, Federal Energy Regulatory Commission, Washington, D.C., 1980.
14. **Golden, J., Ouellette, R. P., Saari, S., and Cheremisinoff, P. N.,** *Environmental Impact Data Book,* Ann Arbor Science, Ann Arbor, Mich., 1980.
15. **Sullivan, T. F. P., Ed.,** *Environmental Statutes,* 1986 ed., Government Institutes, Rockville, Md., 1986.
16. **Hunter, R. L.,** Preliminary Scenarios for the Release of Radioactive Waste from a Hypothetical Repository in Basalt of the Columbia Plateau, NUREG/CR-3353, Nuclear Regulatory Commission, Washington, D.C., 1983.
17. **Jain, R. K., Urban, L. V., and Stacey, G. S.,** *Environmental Impact Analysis,* 2nd ed., Van Nostrand Reinhold, New York, 1981.
18. **Koenigsberger, M. D.,** Preventing pollution at the source, *Chem. Eng. Prog.,* 82, 5, 1986.
19. **Licht, W.,** *Air Pollution Control Engineering, Basic Calculations for Particulate Collection,* Marcel Dekker, New York, 1980.
20. **Metcalf & Eddy, Inc.,** *Wastewater Engineering: Collection, Treatment, Disposal, and Reuse,* 2nd ed., Tchobanogious, G., Ed., New York, 1979.
21. **Moore, E. G.,** Control Technology for Radioactive Emissions to the Atmosphere at U.S. Department of Energy Facilities, U.S. Environmental Protection Agency, Montgomery, Alabama, 1983.
22. **Murthy, K. S., Rosenberg, H. S., and Engdahl, R. B.,** Status and problems of flue gas desulfurization processes, *J. Air Pollut. Control Assoc.,* 26, 9, 1976.

23. **Murthy, K. S.,** Environmental assessment of fluidized-bed combustion systems, *CHEMTECH,* 10, 58, 1980.
24. Review of the Current State of Radiation Protection Philosophy, Rep. No. 43, National Council on Radiation Protection and Measurements (NCRP), Bethesda, Md., 1975.
25. **Ortolano, L.,** Estimating air quality impacts, *Environ. Impact Assess. Rev.,* 5, 1, 1985.
26. **Platt, A. M., Robinson, J. V., and Hill, O. F., Ed.,** *Nuclear Fact Book,* 2nd ed., Harwood Academic, New York, 1985.
27. **Powell, J. A. and Liebetrau, S. F.,** Effective production of very large documents, in *Proc. 34th Int. Technical Communications Conf.,* Society for Technical Communication, Washington, D.C., 1987, WE 210.
28. Database Issue (McGraw-Hill publication, New York), *Power,* 130, 11, 1986.
29. **Quade, E. S.,** *Analysis for Public Decisions: 2nd ed.,* Elsevier New York, 1984.
30. **Rubin, E. S., Cushey, M. A., Marnicio, R. J., Bloyd, C. N., and Skea, J. F.,** Controlling acid deposition: the role of FGD, *Environ. Sci. Technol.,* 20, 10, 1986.
31. **Stern, A. C.,** *Air Pollution,* Vol. 4, 3rd ed., Academic Press, New York, 1977.
32. **Sugden, R. and Williams, A.,** *The Principles of Practical Cost Benefit Analysis,* Oxford University Press, Oxford, 1978.
33. Environmental Compliance Guide, Vol. 1 and 2, DOE/EV-0132, U.S. Department of Energy, Washington, D.C., 1981.
34. Final Environmental Impact Statement, Management of Commercially Generated Radioactive Waste, DOE/EIS-0046F, U.S. Department of Energy, Washington, D.C., 1980.
35. Air Quality Criteria for Sulfur Oxides, Publ. No. AP-50, U.S. Department of Health, Education, and Welfare, Washington, D.C., 1969.
36. *Washington Post (Washington, D.C.),* March 14, 1984.

GLOSSARY

AAQS	Ambient air quality standards
ACFM	Actual cubic feet per minute of gas measured at the actual temperature and pressure of the gas
ACHP	Advisory Committee on Historic Preservation
ADM	Action description memorandum
AQCR	Air quality control region
BOD	Biological oxygen demand
Btu/hr	British thermal units/hour
CAA	Clean Air Act
$CaSO_4$	Calcium sulfate
C & C	Consultation and cooperation
CDM	A climatological steady-state Gaussian plume model to estimate seasonal pollutant concentrations in an urban area
CEQ	Council on Environmental Quality
CERCLA	Comprehensive Environmental Response, Compensation and Liability Act of 1980
CFR	Code of Federal Regulations
CO	Carbon monoxide
COD	Chemical oxygen demand
CRSTER	Air Quality Diffusion Model
CWA	Clean Water Act
°F	Degrees Farenheit
K	Kelvin
DEIS	Draft environmental impact statement
CH_4	Methane
DHEC	Department of Health and Environmental Control
DO	Dissolved oxygen
DOE	U.S. Department of Energy
DOI	U.S. Department of Interior
EA	Environmental assessment
EC	Environmental concerns
EE	Environmental evaluation
EIS	Environmental impact statement
EO	Executive Order
EO	Environmental objections
EPA	U.S. Environmental Protection Agency
ER	Environmental report
ESA	Endangered Species Act
ESP	Electrostatic precipitator
EU	Environmentally unsatisfactory
FR	*Federal Register*
FEIS	Final environmental impact statement
FERC	Federal Energy Regulatory Commission
FGD	Flue gas desulfurization
FONSI	Finding of no significant impact
ft	Foot (feet)
grains (gr)	A unit of weight (7000 grains = 1 pound)
H_2O	Water (a molecular formula for water)
HEPA	High efficiency particulate air filter

HOP	Historic Preservation Office
HUD	U.S. Department of Housing and Urban Development
IP	Implementation plan
km	Kilometers
KWh	Kilowatthour, a unit of electrical energy
lb/hr	Pounds per hour
L/G ratio	Liquid-to-gas ratio
LO	Lack of objections
LEIS	Legislative environmental impact statement
LSA	List of (CFR) Sections Affected, issued monthly
LUST	Leaking underground storage tanks
m	Meters
MRS	Monitored retrievable storage
MTU	Metric tons of uranium
MW	Megawatts electricity
NAA	Nonattainment area
NAAQS	National ambient air quality standards
NEPA	National Environmental Policy Act of 1969
NESHAP	National emission standards for hazardous air pollutants
NO_2	Nitrogen dioxide
NO_x	Oxides of nitrogen
NCA	Noise Control Act
NOA	Notice of availability
NOI	Notice of intent
NPDES	National pollutant discharge elimination system
NRC	U.S. Nuclear Regulatory Commission
NRDC	Natural Resources Defense Council
NSPS	New source performance standard
NWPA	Nuclear Waste Policy Act
OMB	Office of Management and Budget
OPPE	Office of Policy Planning and Evaluation
OSHA	Occupational Safety and Health Act
Pascals	A unit of pressure
%	Percent
pH	A measure of acidity or alkalinity, acidic if pH is <7, neutral if $=7$, and alkaline if >7
PSD	Prevention of significant deterioration
PTMAX	A steady-state Gaussian plume model for calculating ambient air concentration from a single point source
PTMTPW	A steady-state Gaussian plume model
RCRA	Resource Conservation and Recovery Act
RIA	Regulatory impact analysis
ROD	Record of decision
SAR	Start action request
SCC	Steering Committee chairman
SCFM	Standard cubic feet per minute measured at 25°C and 1 atmosphere pressure
SDWA	Safe Drinking Water Act
SEC	Secretary
sec	Second
SIP	State implementation plan

SO_2	Sulfur dioxide
TCA	Turbulent contact absorber
TDS	Total dissolved solids
TOC	Total organic carbon
TOSCA	Toxic Substances Control Act
TSP	Total suspended particulates
UIC	Underground injection control
UOP	Universal Oil Products Company
U.S.C.	United States Code
UTM	Universal Transverse Mercator Coordinates

APPENDIX A

NATIONAL ENVIRONMENTAL POLICY ACT OF 1969, AS AMENDED*

An Act to establish a national policy for the environment, to provide for the establishment of a Council on Environmental Quality, and for other purposes.

Be it enacted by the Senate and House of Representatives of the United States of America in Congress assembled, That this Act may be cited as the ''National Environmental Policy Act of 1969.''

Purpose

Sec. 2. The purposes of this Act are: To declare a national policy which will encourage productive and enjoyable harmony between man and his environment; to promote efforts which will prevent or eliminate damage to the environment and biosphere and stimulate the health and welfare of man; to enrich the understanding of the ecological systems and natural resources important to the Nation; and to establish a Council on Environmental Quality.

TITLE I

Declaration of National Environmental Policy

Sec. 101. (a) The Congress, recognizing the profound impact of man's activity on the interrelations of all components of the natural environment, particularly the profound influences of population growth, high-density urbanization, industrial expansion, resource exploitation, and new and expanding technological advances and recognizing further the critical importance of restoring and maintaining environmental quality to the overall welfare and development of man, declares that it is the continuing policy of the Federal Government, in cooperation with State and local governments, and other concerned public and private organizations, to use all practicable means and measures, including financial and technical assistance, in a manner calculated to foster and promote the general welfare, to create and maintain conditions under which man and nature can exist in productive harmony, and fulfill the social, economic, and other requirements of present and future generations of Americans.

(b) In order to carry out the policy set forth in this Act, it is the continuing responsibility of the Federal Government to use all practicable means, consistent with other essential considerations of national policy, to improve and coordinate Federal plans, functions, programs, and resources to the end that the Nation may —

(1) fulfill the responsibilities of each generation as trustee of the environment for succeeding generations;

(2) assure for all Americans safe, healthful, productive, and esthetically and culturally pleasing surroundings;

(3) attain the widest range of beneficial uses of the environment without degradation, risk to health or safety, or other undesirable and unintended consequences;

(4) preserve important historic, cultural, and natural aspects of our national heritage, and maintain, wherever possible, an environment which supports diversity and variety of individual choice;

* Pub. L. 91-190, 42 U.S.C. 4321-4347, January 1, 1970, as amended by Pub. L. 94-52, July 3, 1975, and Pub. L. 94-83, August 9, 1975.

(5) achieve a balance between population and resource use which will permit high standards of living and a wide sharing of life's amenities; and
(6) enhance the quality of renewable resources and approach the maximum attainable recycling of depletable resources.

(c) The Congress recognizes that each person should enjoy a healthful environment and that each person has a responsibility to contribute to the preservation and enhancement of the environment.

Sec. 102. The Congress authorizes and directs that, to the fullest extent possible: (1) the policies, regulations, and public laws of the United States shall be interpreted and administered in accordance with the policies set forth in this Act, and (2) all agencies of the Federal Government shall —

(A) Utilize a systematic, interdisciplinary approach which will insure the integrated use of the natural and social sciences and the environmental design arts in planning and in decisionmaking which may have an impact on man's environment;
(B) Identify and develop methods and procedures, in consultation with the Council on Environmental Quality established by title II of this Act, which will insure that presently unquantified environmental amenities and values may be given appropriate consideration in decisionmaking along with economic and technical considerations;
(C) Include in every recommendation or report on proposals for legislation and other major Federal actions significantly affecting the quality of the human environment, a detailed statement by the responsible official on —

(i) The environmental impact of the proposed action,
(ii) Any adverse environmental effects which cannot be avoided should the proposal be implemented,
(iii) Alternatives to the proposed action,
(iv) The relationship between local short-term uses of man's environment and the maintenance and enhancement of long-term productivity, and,
(v) Any irreversible and irretrievable commitments of resources which would be involved in the proposed action should it be implemented.

Prior to making any detailed statement, the responsible Federal official shall consult with and obtain the comments of any Federal agency which has jurisdiction by law or special expertise with respect to any environmental impact involved. Copies of such statement and the comments and views of the appropriate Federal, State, and local agencies, which are authorized to develop and enforce environmental standards, shall be made available to the President, the Council on Environmental Quality and to the public as provided by section 552 of Title 5, United States Code, and shall accompany the proposal through the existing agency review processes;
(D) Any detailed statement required under subparagraph (c) after January 1, 1970, for any major Federal action funded under a program of grants to States shall not be deemed to be legally insufficient solely by reason of having been prepared by a State agency or official, if:

(i) the State agency or official has statewide jurisdiction and has the responsibilty for such action,
(ii) the responsible Federal official furnishes guidance and participates in such preparation,
(iii) the responsible Federal official independently evaluates such statement prior to its approval and adoption, and

(iv) after January 1, 1976, the responsible Federal official provides early notification to, and solicits the view of, any other State or any Federal land management entity of any action or any alternative thereto which may have significant impacts upon such State or affected Federal land management entity and, if there is any disagreement on such impacts, prepares a written assessment of such impacts and views for incorporation into such detailed statement.

The procedures in this subparagraph shall not relieve the Federal official of his responsibilities for the scope, objectivity, and content of the entire statement or of any other responsibility under this Act; and further, this subparagraph does not affect the legal sufficiency of statements prepared by state agencies with less than statewide jurisdiction.

(E) Study, develop, and describe appropriate alternatives to recommend course of action in any proposal which involves unresolved conflicts concerning alternative uses of available resources;

(F) Recognize the worldwide and long-range character of environmental problems and, where consistent with the foreign policy of the United States, lend appropriate support to initiatives, resolutions, and programs designed to maximize international cooperation in anticipating and preventing a decline in the quality of mankind's world environment;

(G) Make available to States, counties, municipalities, institutions, and individuals, advice and information useful in restoring, maintaining, and enhancing the quality of the environment;

(H) Initiate and utilize ecological information in the planning and development of resource-oriented projects; and

(I) Assist the Council on Environmental Quality established by title II of this Act.

Sec. 103. All agencies of the Federal Government shall review their present statutory authority, administrative regulations, and current policies and procedures for the purpose of determining whether there are any deficiencies or inconsistencies therein which prohibit full compliance with the purposes and provides of this Act and shall propose to the President not later than July 1, 1971, such measures as may be necessary to bring their authority and policies into conformity with the intent, purposes, and procedures set forth in this Act.

Sec. 104. Nothing in section 102 or 103 shall in any way affect the specific statutory obligations of any Federal agency (1) to comply with criteria or standards of environmental quality, (2) to coordinate or consult with any other Federal or State agency, or (3) to act, or refrain from acting contingent upon the recommendations or certification of any other Federal or State agency.

Sec. 105. The policies and goals set forth in this Act are supplementary to those set forth in existing authorizations of Federal agencies.

TITLE II

Council on Environmental Quality

Sec. 201. The President shall transmit to the Congress annually beginning July 1, 1970, an Environmental Quality Report (hereinafter referred to as the ''report'') which shall set forth (1) the status and condition of the major natural, manmade, or altered environmental classes of the Nation, including, but not limited to, the air, the aquatic, including marine, estuarine, and fresh water, and the terrestrial environment, including, but not limited to, the forest, dryland, wetland, range, urban, suburban, and rural environment; (2) current and foreseeable trends in the quality, managment, and utliziation of such environments and the effects of those trends on the social, economic, and other requirements of the Nation; (3)

the adequacy of available natural resources for fulfilling human and economic requirements of the Nation in the light of expected population pressures; (4) a review of the programs and activities (including regulatory activities) of the Federal Government, the State and local governments, and nongovernmental entities or individuals, with particular reference to their effect on the environment and on the conservation, development and utilization of natural resources; and (5) a program for remedying the deficiencies of existing programs and activities, together with recommendations for legislation.

Sec. 202. There is created in the Executive Office of the President a Council on Environmental Quality (hereinafter referred to as the "Council"). The Council shall be composed of three members who shall be appointed by the President to serve at his pleasure, by and with the advice and consent of the Senate. The President shall designate one of the members of the Council to serve as Chairman. Each member shall be a person who, as a result of his training, experience, and attainments, is exceptionally well qualified to analyze and interpret environmental trends and information of all kinds; to appraise programs and activities of the Federal Government in the light of the policy set forth in title I of this Act; to be conscious of and responsive to the scientific, economic, social, esthetic, and cultural needs and interests of the Nation; and to formulate and recommend national policies to promote the improvement of the quality of the environment.

Sec. 203. The Council may employ such officers and employees as may be necessary to carry out its functions under this Act. In addition, the Council may employ and fix the compensation of such experts and consultants as may be necessary for the carrying out of its functions under this Act, in accordance with section 3109 of title 5, United States Code (but without regard to the last sentence thereof).

Sec. 204. It shall be the duty and function of the Council —

(1) to assist and advise the President in the preparation of the Environmental Quality Report required by section 201 of this title;
(2) to gather timely and authoritative information concerning the conditions and trends in the quality of the environment both current and prospective, to analyze and interpret such information for the purpose of determining whether such conditions and trends are interfering, or are likely to interfere, with the achievement of the policy set forth in title I of this Act, and to compile and submit to the President studies relating to such conditions and trends;
(3) to review and appraise the various programs and activities of the Federal Government in the light of the policy set forth in title I of this Act for the purpose of determining the extent to which such programs and activities are contributing to the achievement of such policy, and to make recommendations to the President with respect thereto;
(4) to develop and recommend to the President national policies to foster and promote the improvement of environmental quality to meet the conservation, social, economic, health, and other requirements and goals of the Nation;
(5) to conduct investigations, studies, surveys, research, and analyses relating to ecological systems and environmental quality;
(6) to document and define changes in the natural environment, including the plant and animal systems, and to accumulate necessary data and other information for a continuing analysis of these changes or trends and an interpretation of their underlying causes;
(7) to report at least once each year to the President on the state and condition of the environment; and
(8) to make and furnish such studies, reports thereon, and recommendations with respect to matters of policy and legislation as the President may request.

Sec. 205. In exercising its powers, functions, and duties under this Act, the Council shall —

(1) Consult with the Citizens' Advisory Committee on Environmental Quality established by Executive Order numbers 11472, dated May 29, 1969, and with such representatives of science, industry, agriculture, labor, conservation organizations, State and local governments and other groups, as it deems advisable; and

(2) Utilize, to the fullest extent possible, the services, facilities, and information (including statistical information) of public and private agencies and organizations, and individuals, in order that duplication of effort and expense may be avoided, thus assuring that the Council's activities will not unnecessarily overlap or conflict with similar activities authorized by law and performed by established agencies.

Sec. 206. Members of the Council shall serve full time and the Chairman of the Council shall be compensated at the rate provided for Level II of the Executive Schedule Pay Rates (5 U.S.C. 5313). The other members of the Council shall be compensated at the rate provided for Level IV of the Executive Schedule Pay Rates (5 U.S.C. 5315).

Sec. 207. The Council may accept reimbursements from any private nonprofit organization or from any department, agency, or instrumentality of the Federal Government, any State, or local government, for the reasonable travel expenses incurred by an officer or employee of the Council in connection with his attendance at any conference, seminar, or similar meeting conducted for the benefit of the Council.

Sec. 208. The Council may make expenditures in support of its international activities, including expenditures for: (1) international travel; (2) activities in implementation of international agreements; and (3) the support of international exchange programs in the United States and in foreign countries.

Sec. 209. There are authorized to be appropriated to carry out the provisions of this chapter not to exceed $300,000 for fiscal year 1970, $700,000 for fiscal year 1971, and $1,000,000 for each fiscal year thereafter.

APPENDIX B

CONTENTS

For further information, contact:
General Counsel
Council on Environmental Quality
Executive Office of the President
722 Jackson Place, N.W.
Washington, D.C. 20006
(202)395-5754

Council on Environmental Quality
Executive Office of the President

REGULATIONS
For Implementing the Procedural Provisions of the
NATIONAL ENVIRONMENTAL POLICY ACT

Reprint
40 CFR Parts 1500-1508
(as of July 1, 1986)

TABLE OF CONTENTS

PART 1500 — PURPOSE, POLICY, AND MANDATE

PART 1501 — NEPA AND AGENCY PLANNING

PART 1502 — ENVIRONMENTAL IMPACT STATEMENT

PART 1503 — COMMENTING

PART 1504 — PREDECISION REFERRALS TO THE COUNCIL OF PROPOSED FEDERAL ACTIONS DETERMINED TO BE ENVIRONMENTALLY UNSATISFACTORY

PART 1505 — NEPA AND AGENCY DECISIONMAKING

PART 1506 — OTHER REQUIREMENTS OF NEPA

PART 1507 — AGENCY COMPLIANCE

PART 1508 — TERMINOLOGY AND INDEX

PART 1500 — PURPOSE, POLICY, AND MANDATE

Sec.

Authority: NEPA, the Environmental Quality Improvement Act of 1970, as amended (42 U.S.C. 4371 *et seq.*), sec. 309 of the Clean Air Act, as amended (42 U.S.C. 7609) and Executive Order 11514 (March 5, 1970 as amended by Executive Order 11991, May 24, 1977).

Source: 43 FR 55990, Nov. 29, 1978, unless otherwise noted.

§1500.1 Purpose.

(a) The National Environmental Policy Act (NEPA) is our basic national charter for protection of the environment. It establishes policy, sets goals (section 101), and provides means (section 102) for carrying out the policy. Section 102(2) contains ''action-forcing'' provisions to make sure that federal agencies act according to the letter and spirit of the Act. The regulations that follow implement section 102(2). Their purpose is to tell federal agencies what they must do to comply with the procedures and achieve the goals of the Act. The President, the federal agencies, and the courts share responsibility for enforcing the Act so as to achieve the substantive requirements of section 101.

(b) NEPA procedures must insure that environmental information is available to public officials and citizens before decisions are made and before actions are taken. The information must be of high quality. Accurate scientific analysis, expert agency comments, and public scrutiny are essential to implementing NEPA. Most important, NEPA documents must concentrate on the issues that are truly significant to the action in question, rather than amassing needless detail.

(c) Ultimately, of course, it is not better documents but better decisions that count. NEPA's purpose is not to generate paperwork — even excellent paperwork — but to foster excellent action. The NEPA process is intended to help public officials make decisions that are based on understanding of environmental consequences, and take actions that protect, restore, and enhance the environment. These regulations provide the direction to achieve this purpose.

§ 1500.2 Policy

Federal agencies shall to the fullest extent possible:

(a) Interpret and administer the policies, regulations, and public laws of the United States in accordance with the policies set forth in the Act and in these regulations.

(b) Implement procedures to make the NEPA process more useful to decisionmakers and the public; to reduce paperwork and the accumulation of extraneous background data; and to emphasize real environmental issues and alternatives. Environmental impact statements shall be concise, clear, and to the point, and shall be supported by evidence that agencies have made the necessary environmental analyses.

(c) Integrate the requirements of NEPA with other planning and environmental review procedures required by law or by agency practice so that all such procedures run concurrently rather than consecutively.

(d) Encourage and facilitate public involvement in decisions which affect the quality of the human environment.

(e) Use the NEPA process to identify and assess the reasonable alternatives to proposed actions that will avoid or minimize adverse effects of these actions upon the quality of the human environment.

(f) Use all practicable means, consistent with the requirements of the Act and other essential considerations of national policy, to restore and enhance the quality of the human environment and avoid or minimize any possible adverse effects of their actions upon the quality of the human environment.

§ 1500.3 Mandate.

Parts 1500 through 1508 of this title provide regulations applicable to and binding on all Federal agencies for implementing the procedural provisions of the National Environmental Policy Act of 1969, as amended (Pub. L. 91-190, 42 U.S.C. 4321 et seq.) (NEPA or the Act) except where compliance would be inconsistent with other statutory requirements. These regulations are issued pursuant to NEPA, the Environmental Quality Improvement Act of 1970, as amended (42 U.S.C. 4371 et seq.) Section 309 of the Clean Air Act, as amended (42 U.S.C. 7609) and Executive Order 11514, Protection and Enhancement of Environmental Quality (March 5, 1970, as amended by Executive Order 11991, May 24, 1977). These regulations, unlike the predecessor guidelines, are not confined to sec. 102(2)(C) (environmental impact statements). The regulations apply to the whole of section 102(2). The provisions of the Act and of these regulations must be read together as a whole in order to comply with the spirit and letter of the law. It is the Council's intention that judicial review of agency compliance with these regulations not occur before an agency has filed the final environmental impact statement, or has made a final finding of no significant impact (when such a finding will result in action affecting the environment), or takes action that will result in irreparable injury. Furthermore, it is the Council's intention that any trivial violation of these regulations not give rise to any independent cause of action.

§ 1500.4 Reducing paperwork.

Agencies shall reduce excessive paperwork by:

(a) Reducing the length of environmental impact statements (§ 1502.2(c)), by means such as setting appropriate page limits (§§ 1501.7(b)(1) and 1502.7).

(b) Preparing analytic rather than encyclopedic environmental impact statements (§ 1502.2(a)).

(c) Discussing only briefly issues other than significant ones (§ 1502.2(b)).

(d) Writing environmental impact statements in plain language (§ 1502.8).

(e) Following a clear format for environmental impact statements (§ 1502.10).

(f) Emphasizing the portions of the environmental impact statement that are useful to decisionmakers and the public (§§ 1502.14 and 1502.15) and reducing emphasis on background material (§ 1502.16).

(g) Using the scoping process, not only to identify significant environmental issues deserving of study, but also to deemphasize insignificant issues, narrowing the scope of the environmental impact statement process accordingly (§ 1501.7).

(h) Summarizing the environmental impact statement (§ 1502.12) and circulating the summary instead of the entire environmental impact statement if the latter is unusually long (§ 1502.19).

(i) Using program, policy, or plan environmental impact statements and tiering from statements of broad scope to those of narrower scope, to eliminate repetitive discussions of the same issues (§§ 1502.4 and 1502.20).

(j) Incorporating by reference (§ 1502.21).

(k) Integrating NEPA requirements with other environmental review and consultation requirements (§ 1502.25).

(l) Requiring comments to be as specific as possible (§ 1503.3).

(m) Attaching and circulating only changes to the draft environmenal impact statement, rather than rewriting and circulating the entire statement when changes are minor (§ 1503.4(c)).

(n) Eliminating duplication with State and local procedures, by providing for joint preparation (§ 1506.2), and with other Federal procedures, by providing that an agency may adopt appropriate environmental documents prepared by another agency (§ 1506.3).

(o) Combining environmental documents with other documents (§ 1506.4).

(p) Using categorical exclusions to define categories of actions which do not individually or cumulatively have a significant effect on the human environment and which are therefore exempt from requirements to prepare an environmental impact statement (§ 1508.4).

(q) Using a finding of no significant impact when an action not otherwise excluded will not have a significant effect on the human environment and is therefore exempt from requirements to prepare an environmental impact statement (§ 1508.13).

[43 FR 55990, Nov. 29, 1978; 44 FR873, Jan. 3, 1979]

§ 1500.5 Reducing delay.

Agencies shall reduce delay by:

(a) Integrating the NEPA process into early planning (§ 1501.2).

(b) Emphasizing interagency cooperation before the environmental impact statement is prepared, rather than submission of adversary components on a completed document (§ 1501.6).

(c) Insuring the swift and fair resolution of lead agency disputes (§ 1501.5).

(d) Using the scoping process for an early identification of what are and what are not the real issues (§ 1501.7).

(e) Establishing appropriate time limits for the environmental impact statement process (§§ 1501.7(b)(2) and 1501.8).

(f) Preparing environmental impact statements early in the process (§ 1502.5).

(g) Integrating NEPA requirements with other environmental review and consultation requirements (§ 1502.25).

(h) Eliminating duplication with State and local procedures by providing for joint preparation (§ 1506.2) and with other Federal procedures by providing that an agency may adopt appropriate environmental documents prepared by another agency (§ 1506.3).

(i) Combining environmental documents with other documents (§ 1506.4).

(j) Using accelerated procedures for proposals for legislation (§ 1506.8).

(k) Using categorical exclusions to define categories of actions which do not individually or cumulatively have a significant effect on the human environment (§ 1508.4) and which are therefore exempt from requirements to prepare an environmental impact statement.

(l) Using a finding of no significant impact when an action not otherwise excluded will not have a significant effect on the human environment (§ 1508.13) and is therefore exempt from requirements to prepare an environmental impact statement.

§ 1500.6 Agency authority.

Each agency shall interpret the provisions of the Act as a supplement to its existing authority and as a mandate to view traditional policies and missions in the light of the Act's national environmental objectives. Agencies shall review their policies, procedures, and regulations accordingly and revise them as necessary to insure full compliance with the purposes and provisions of the Act. The phrase "to the fullest extent possible" in section 102 means that each agency of the Federal Government shall comply with that section unless existing law applicable to the agency's operations expressly prohibits or makes compliance impossible.

PART 1501 — NEPA AND AGENCY PLANNING

Authority: NEPA, the Environmental Quality Improvement Act of 1970, as amended (42 U.S.C. 4371 *et seq.*), sec. 309 of the Clean Air Act, as amended (42 U.S.C. 7609) and Executive Order 11514 (March 5, 1970, as amended by Executive Order 11991, May 24, 1977).

Source: 43 FR 55992, Nov. 29, 1978, unless otherwise noted.

§ 1501.1 Purpose.

The purposes of this part include:

(a) Integrating the NEPA process into early planning to insure appropriate consideration of NEPA's policies and to eliminate delay.

(b) Emphasizing cooperative consultation among agencies before the environmental impact statement is prepared rather than submission of adversary comments on a completed document.

(c) Providing for the swift and fair resolution of lead agency disputes.

(d) Identifying at an early stage the significant environmental issues deserving of study and deemphasizing insignificant issues, narrowing the scope of the environmental impact statement accordingly.

(e) Providing a mechanism for putting appropriate time limits on the environmental impact statement process.

§ 1501.2 Apply NEPA early in the process.

Agencies shall integrate the NEPA process with other planning at the earliest possible time to insure that planning and decisions reflect environmental values, to avoid delays later in the process, and to head off potential conflicts. Each agency shall:

(a) Comply with the mandate of section 102(2)(A) to "utilize a systematic, interdisciplinary approach which will insure the integrated use of the natural and social sciences and the environmental design arts in planning and in decisionmaking which may have an impact on man's environment," as specified by § 1507.2.

(b) Identify environmental effects and values in adequate detail so they can be compared to economic and technical analyses. Environmental documents and appropriate analyses shall be circulated and reviewed at the same time as other planning documents.

(c) Study, develop, and describe appropriate alternatives to recommend courses of action in any proposal which involves unresolved conflicts concerning alternative uses of available resources as provided by section 102(2)(E) of the Act.

(d) Provide for cases where actions are planned by private applicants or other non-Federal entities before Federal involvement so that:

(1) Policies or designated staff are available to advise potential applicants of studies or other information foreseeably required for later Federal Action.

(2) The Federal agency consults early with appropriate State and local agencies and Indian tribes and with interested private persons and organizations when its own involvement is reasonably foreseeable.

(3) The Federal agency commences its NEPA process at the earliest possible time.

§ 1501.3 When to prepare an environmental assessment.

(a) Agencies shall prepare an environmental assessment (§ 1508.9) when necessary under the procedures adopted by individual agencies to supplement these regulations as described in § 1507.3. An assessment is not necessary if the agency has decided to prepare an environmental impact statement.

(b) Agencies may prepare an environmental assessment on any action at any time in order to assist agency planning and decisionmaking.

§ 1501.4 Whether to prepare an environmental impact statement.

In determining whether to prepare an environmental impact statement the Federal agency shall:

(a) Determine under its procedures supplementing these regulations (described in § 1507.3) whether the proposal is one which:

(1) Normally requires an environmental impact statement, or

(2) Normally does not require either an environmental impact statement or an environmental assessment (categorical exclusion).

(b) If the proposed action is not covered by paragraph (a) of this section, prepare an environmental assessment (§ 1508.9). The agency shall involve environmental agencies, applicants, and the public, to the extent practicable, in preparing assessments required by § 1508.9(a)(1).

(c) Based on the environmental assessment make its determination whether to prepare an environmental impact statement.

(d) Commence the scoping process (§ 1501.7), if the agency will prepare an environmental impact statement.

(e) Prepare a finding of no significant impact (§ 1508.13), if the agency determines on the basis of the environmental assessment not to prepare a statement.

(1) The agency shall make the finding of no significant impact available to the affected public as specified in § 1506.6.

(2) In certain limited circumstances, which the agency may cover in its procedures under § 1507.3, the agency shall make the finding of no significant impact available for public review (including State and areawide clearing houses) for 30 days before the agency makes its final determination whether to prepare an environmental impact statement and before the action may begin. The circumstances are:

(i) The proposed action is, or is closely similar to, one which normally requires the preparation of an environmental impact statement under the procedures adopted by the agency pursuant to § 1507.3, or

(ii) The nature of the proposed action is one without precedent.

§ 1501.5 Lead agencies.

(a) A lead agency shall supervise the preparation of an environmental impact statement if more than one Federal agency either:

(1) Proposes or is involved in the same action; or

(2) Is involved in a group of actions directly related to each other because of their functional interdependence or geographical proximity.

(b) Federal, State, or local agencies, including at least one Federal agency, may act as joint lead agencies to prepare an environmental impact statement (§ 1506.2).

(c) If an action falls within the provisions of paragraph (a) of this section the potential lead agencies shall determine by letter or memorandum which agency shall be the lead agency and which shall be the lead agency and which shall be cooperating agencies. The agencies shall resolve the lead agency question so as not to cause delay. If there is disagreement among the agencies, the following factors (which are listed in order of descending importance) shall determine lead agency designation:

(1) Magnitude of agency's involvement.

(2) Project approval/disapproval authority.

(3) Expertise concerning the action's environmental effects.

(4) Duration of agency's involvement.

(5) Sequence of agency's involvement.

(d) Any Federal agency, or any State or local agency or private person substantially affected by the absence of lead agency designation, may make a written request to the potential lead agencies that a lead agency be designated.

(e) If Federal agencies are unable to agree on which agency will be the lead agency or if the procedure described in paragraph (c) of this section has not resulted within 45 days in a lead agency designation, any of the agencies or persons concerned may file a request with the Council asking it to determine which Federal agency shall be the lead agency.

A copy of the request shall be transmitted to each potential lead agency. The request shall consist of:

(1) A precise description of the nature and extent of the proposed action.

(2) A detailed statement of why each potential lead agency should or should not be the lead agency under the criteria specified in paragraph (c) of this section.

(f) A response may be filed by any potential lead agency concerned within 20 days after a request is filed with the Council. The Council shall determine as soon as possible but not later than 20 days after receiving the request and all responses to it which Federal agency shall be the lead agency and which other Federal agencies shall be cooperating agencies.

[43 FR 55992, Nov. 29, 1978; 44 FR 873, Jan. 3, 1979]

§ 1501.6 Cooperating agencies.

The purpose of this section is to emphasize agency cooperation early in the NEPA process. Upon request of the lead agency, any other Federal agency which has jurisdiction by law shall be a cooperating agency. In addition any other Federal agency which has special expertise with respect to any environmental issue, which should be addressed in the statement may be a cooperating agency upon request of the lead agency. An agency may request the lead agency to designate it a cooperating agency.

(a) The lead agency shall:

(1) Request the participation of each cooperating agency in the NEPA process at the earliest possible time.

(2) Use the environmental analysis and proposals of cooperating agencies with jurisdiction by law or special expertise, to the maximum extent possible consistent with its responsibility as lead agency.

(3) Meet with a cooperating agency at the latter's request.

(b) Each cooperating agency shall:

(1) Participate in the NEPA process at the earliest possible time.

(2) Participate in the scoping process (described below in § 1501.7).

(3) Assume on request of the lead agency responsibility for developing information and preparing environmental analyses including portions of the environmental impact statement concerning which the cooperating agency has special expertise.

(4) Make available staff support at the lead agency's request to enhance the latter's interdisciplinary capability.

(5) Normally use its own funds. The lead agency shall, to the extent available funds permit, fund those major activities or analyses it request from cooperating agencies. Potential lead agencies shall include such funding requirements in their budget request.

(c) A cooperating agency may in response to a lead agency's request for assistance in preparing the environmental impact statement (described in paragraph (b) (3), (4), or (5) of this section) reply that other program commitments preclude any involvement or the degree of involvement request in the action that is the subject of the environmental impact statement. A copy of this reply shall be submitted to the Council.

§ 1501.7 Scoping.

There shall be an early and open process for determining the scope of issues to be addressed and for identifying the significant issues related to a proposed action. This process shall be termed scoping. As soon as practicable after its decision to prepare environmental impact statement and before the scoping process the lead agency shall publish a notice of intent (§ 1508.22) in the *Federal Register* except as provided in § 1507.3(e).

(a) As part of the scoping process the lead agency shall:

(1) Invite the participation of affected Federal, State, and local agencies, any affected Indian tribe, the proponent of the action, and other interested persons (including those who might not be in accord with the action on environmental grounds), unless there is a limited exception under § 1507.3(c). An agency may give notice in accordance with § 1506.6.

(2) Determine the scope (§ 1508.25) and the significant issues to be analyzed in depth in the environmental impact statement.

(3) Identify and eliminate from detailed study the issues which are not significant or which have been covered by prior environmental review (§ 1506.3), narrowing the discussion of these issues in the statement to a brief presentation of why they will not have a significant effect on the human environment or providing a reference to their coverage elsewhere.

(4) Allocate assignments for preparation of the environmental impact statement among the lead and cooperative agencies, with the lead agency retaining responsibility for the statement.

(5) Indicate any public environmental assessments and other environmental impact statements which are being or will be prepared that are related to but are not part of the scope of the impact statement under consideration.

(6) Identify other environmental review and consultation requirements so the lead and cooperating agencies may prepare other required analyses and studies concurrently with, and integrated with, the environmental impact statement as provided in § 1502.25.

(7) Indicate the relationship between the timing of the preparation of environmental analyses and the agency's tentative planning and decisionmaking schedule.

(b) As part of the scoping process the lead agency may:

(1) Set page limits on environmental documents (§ 1502.7).

(2) Set time limits (§ 1501.8).

(3) Adopt procedures under § 1507.3 to combine its environmental assessment process with its scoping process.

(4) Hold an early scoping meeting or meetings which may be integrated with any other early planning meeting the agency has. Such a scoping meeting will often be appropriate when the impacts of a particular action are confined to specific sites.

(c) An agency shall revise the determinations made under paragraphs (a) and (b) of this section if substantial changes are made later in the proposed action, or if significant new circumstances or information arise which bear on the proposal or its impacts.

§ 1501.8 Time limits.

Although the Council has decided that prescribed universal time limits for the entire NEPA process are too inflexible, Federal agencies are encouraged to set time limits appropriate to individual actions (consistent with the time intervals required by § 1506.10). When multiple agencies are involved the reference to agency below means lead agency.

(a) The agency shall set time limits if an applicant for the proposed action requests them: Provided, that the limits are consistent with the purposes of NEPA and other essential considerations of national policy.

(b) The agency may:

(1) Consider the following factors in determining time limits:

(i) Potential for environmental harm.

(ii) Size of the proposed action.

(iii) State of the art of analytic techniques.

(iv) Degree of public need for the proposed action, including the consequences of delay.

(v) Number of persons and agencies affected.

(vi) Degree to which relevant information is known and if not known the time required for obtaining it.

(vii) Other time limits imposed on the agency by law, regulations, or executive order.

(2) Set overall time limits or limits for each constituent part of the NEPA process, which may include:

(i) Decision on whether to prepare an environmental impact statement (if not already decided).

(ii) Determination of the scope of the environmental impact statement.

(iii) Preparation of the draft environmental impact statement.

(iv) Review of any comments on the draft environmental impact statement from the public and agencies.

(v) Preparation of the final environmental impact statement.

(vi) Review of any comments on the final environmental impact statement.

(vii) Decision on the action based in part on the environmental impact statement.

(3) Designate a person (such as the project manager or a person in the agency's office with NEPA responsibilities) to expedite the NEPA process.

(c) State or local agencies or members of the public may request a Federal Agency to set time limits.

PART 1502 — ENVIRONMENTAL IMPACT STATEMENT

Authority: NEPA, the Environmental Quality Improvement Act of 1970, as amended (42 U.S.C. 4371 *et seq.*), sec. 309 of the Clean Air Act, as amended (42 U.S.C. 7609), and Executive Order 11514 (March 5, 1970, as amended by Executive Order 11991, May 24, 1977).

Source: 43 FR 55994, Nov. 29, 1978, unless otherwise noted.

§ 1502.1 Purpose.

The primary purpose of an environmental impact statement is to serve as an action-forcing device to insure that the policies and goals defined in the Act are infused into the ongoing programs and actions of the Federal Government. It shall provide full and fair discussion of significant environmental impacts and shall inform decision makers and the public of the reasonable alternatives which would avoid or minimize adverse impacts or enhance the quality of the human environment. Agencies shall focus on significant environmental issues and alternatives and shall reduce paperwork and the accumulation of extraneous background data. Statements shall be concise, clear, and to the point, and shall be supported by evidence that the agency has made the necessary environmental analyses. An environmental impact statement is more than a disclosure document. It shall be used by Federal officials in conjunction with other relevant material to plan actions and make decisions.

§ 1502.2 Implementation.

To achieve the purposes set forth in § 1502.1 agencies shall prepare environmental impact statements in the following manner:

(a) Environmental impact statements shall be analytic rather than encyclopedic.

(b) Impacts shall be discussed in proportion to their significance. There shall be only brief discussion of other than significant issues. As in a finding of no significant impact, there should be only enough discussion to show why more study is not warranted.

(c) Environmental impact statements shall be kept concise and shall be no longer than absolutely necessary to comply with NEPA and with these regulations. Length should vary first with potential environmental problems and then with project size.

(d) Environmental impact statements shall state how alternatives considered in it and decisions based on it will or will not achieve the requirements of sections 101 and 102(1) of the Act and other environmental laws and policies.

(e) The range of alternatives discussed in environmental impact statements shall encompass those to be considered by the ultimate agency decisionmaker.

(f) Agencies shall not commit resources prejudicing selection of alternatives before making a final decision (§ 1506.1).

(g) Environmental impact statements shall serve as the means of assessing the environmental impact of proposed agency actions, rather than justifying decisions already made.

§ 1502.3 Statutory requirements for statements.

As required by sec. 102(2)(C) of NEPA environmental impact statements (§ 1508.11) are to be included in every recommendation or report

On proposals (§ 1508.23).
For legislation and (§ 1508.17).
Other major Federal actions (§ 1508.18).
Significantly (§ 1508.27).
Affecting (§§ 1508.3, 1508.8).
The quality of the human environment (§ 1508.14).

§ 1502.4 Major Federal actions requiring the preparation of environmental impact statements.

(a) Agencies shall make sure the proposal which is the subject of an environmental impact statement is properly defined. Agencies shall use the criteria for scope (§ 1508.25) to determine which proposal(s) which are related to each other closely enough to be, in effect, a single course of action shall be evaluated in a single impact statement.

(b) Environmental impact statements may be prepared, and are sometimes required, for broad Federal actions such as the adoption of new agency programs or regulations (§ 1508.18). Agencies shall prepare statements on broad actions so that they are relevant to policy and are timed to coincide with meaningful points in agency planning and decisionmaking.

(c) When preparing statements on broad actions (including proposals by more than one agency), agencies may find it useful to evaluate the proposal(s) in one of the following ways:

(1) Geographically, including actions occurring in the same general location, such as body of water, region, or metropolitan area.

(2) Generically, including actions which have relevant similarities, such as common timing, impacts, alternatives, methods of implementation, media, or subject matter.

(3) By stage of techological development including federal or federally assisted research, development or demonstration programs for new technologies which, if applied, could significantly affect the quality of the human environment. Statements shall be prepared on such programs and shall be available before the program has reached a stage of investment or commitment to implementation likely to determine subsequent development or restrict later alternatives.

(d) Agencies shall as appropriate employ scoping (§ 1501.7), tiering (§ 1502.20), and other methods listed in §§ 1500.4 and 1500.5 to relate broad and narrow actions and to avoid duplication and delay.

§ 1502.5 Timing.

An agency shall commence preparation of an environmental impact statement as close as possible to the time the agency is developing or is presented with a proposal (§ 1508.23) so that preparation can be completed in time for the final statement to be included in any recommendation or report on the proposal. The statement shall be prepared early enough so that it can serve practically as an important contribution to the decisionmaking process and will not be used to rationalize or justify decisions already made (§§ 1500.2(c), 1501.2, and 1502.2). For instance:

(a) For projects directly undertaken by Federal agencies the environmental impact statement shall be prepared at the feasibility analysis (go-no go) stage and may be supplemented at a later stage if necessary.

(b) For applications to the agency appropriate environmental assessments or statements shall be commenced no later than immediately after the application is received. Federal agencies are encouraged to begin preparation of such assessments or statements earlier, preferably jointly with applicable State or local agencies.

(c) For adjudication, the final environmental impact statement shall normally precede the final staff recommendation and that portion of the public hearing related to the impact study. In appropriate circumstances the statement may follow preliminary hearings designed to gather information for use in the statements.

(d) For informal rulemaking the draft environmental impact statement shall normally accompany the proposed rule.

§ 1502.6 Interdisciplinary preparation.

Environmental impact statements shall be prepared using an interdisciplinary approach which will insure the integrated use of the natural and social sciences and the environmental design arts (section 102(2)(A) of the Act). The disciplines of the preparers shall be appropriate to the scope and issues identified in the scoping process (§ 1501.7).

§ 1502.7 Page limits.

The text of final environmental impact statements (e.g., paragraphs (d) through (g) of § 1502.10) shall normally be less than 150 pages and for proposals of unusual scope or complexity shall normally be less than 300 pages.

§ 1502.8 Writing.

Environmental impacts statements shall be written in plain language and may use appropriate graphics so that decisionmakers and the public can readily understand them. Agencies should employ writers of clear prose or editors to write, review, or edit statements, which will be based upon the analysis and supporting data from the natural and social sciences and the environmental design arts.

§ 1502.9 Draft, final, and supplemental statements.

Except for proposals for legislation as provided in § 1506.8 environmental impacts statements shall be prepared in two stages and may be supplemented.

(a) Draft environmental impacts statements shall be prepared in accordance with the scope decided upon in the scoping process. The lead agency shall work with the cooperating agencies and shall obtain comments as required in Part 1503 of this chapter. The draft statement must fulfill and satisfy to the fullest extent possible the requirements established for final statements in section 102(2)(C) of the Act. If a draft statement is so inadequate as to preclude meaningful analysis, the agency shall prepare and circulate a revised draft of the appropriate portion. The agency shall make every effort to disclose and discuss at appropriate points in the draft statement all major points of view on the environmental impacts of the alternatives including the proposed action.

(b) Final environmental impact statements shall respond to comments as required in Part 1503 of this chapter. The agency shall discuss at appropriate points in the final statement any responsible opposing view which was not adequately discussed in the draft statement and shall indicate the agency's response to the issues raised.

(c) Agencies:

(1) Shall prepare supplements to either draft or final environmental impact statements if:

(i) The agency makes substantial changes in the proposed action that are relevant to environmental concerns; or

(ii) There are significant new circumstances or information relevant to environmental concerns and bearing on the proposed action or its impacts.

(2) May also prepare supplements when the agency determines that the purposes of the Act will be furthered by doing so.

(3) Shall adopt procedures for introducing a supplement into its formal administrative record, if such a record exists.

(4) Shall prepare, circulate, and file a supplement to a statement in the same fashion (exclusive of scoping) as a draft and final statement unless alternative procedures are approved by the Council.

§ 1502.10 Recommended format.

Agencies shall use a format for environmental impact statements which will encourage good analysis and clear presentation of the alternatives including the proposed action. The following standard format for environmental impact statements should be followed unless the agency determines that there is a compelling reason to do otherwise:

(a) Cover sheet.
(b) Summary.
(c) Table of contents.
(d) Purpose of and need for action
(e) Alternatives including proposed action (sections 102(2)(C)(iii) and 102(2)(E) of the Act).
(f) Affected environment.
(g) Environmental consequences (especially sections 102(2)(C)(i), (ii), (iv), and (v) of the Act).

(h) List of preparers.
(i) List of Agencies, Organizations, and persons to whom copies of the statement are sent.
(j) Index.
(k) Appendices (if any).

If a different format is used, it shall include paragraphs (a), (b), (c), (h), (i), and (j) of this section and shall include the substance of paragraphs (d), (e), (f), (g), and (k) of this section, as further described in §§ 1502.11 through 1502.18, in any appropriate format.

§ 1502.11 Cover sheet.

The cover sheet shall not exceed one page. It shall include:

(a) A list of the responsible agencies including the lead agency and any cooperating agencies.

(b) The title of the proposed action that is the subject of the statement (and if appropriate the titles of related cooperating agency actions), together with the State(s) and county(ies) (or other jurisdiction if applicable) where the action is located.

(c) The name, address, and telephone number of the person at the agency who can supply further formation.

(d) A designation of the statement as a draft, final, or draft or final supplement.

(e) A one paragraph abstract of the statement.

(f) The date by which comments must be received (computed in cooperation with EPA under § 1506.10).

The information required by this section may be entered on Standard Form 424 (in items 4, 6, 7, 10, and 18).

§ 1052.12 Summary.

Each environmental impact statement shall contain a summary which adequately and accurately summarizes the statement. The summary shall stress the major conclusions, areas of controversy (including issues raised by agencies and the public), and the issues to be resolved (including the choice among alternatives). The summary will normally not exceed 15 pages.

§ 1502.13 Purpose and need.

The statement shall briefly specify the underlying purpose and need to which the agency is responding in proposing the alternatives including the proposed action.

§ 1502.14 Alternatives including the proposed action.

This section is the heart of the environmental impact statement. Based on the information and analysis presented in the sections on the Affected Environment (§ 1502.15) and the Environmental Consequences (§ 1502.16), it should present the environmental impacts of the proposal and the alternatives in comparative form, thus sharply defining the issues and providing a clear basis for choice among options by the decisionmaker and the public. In this section agencies shall:

(a) Rigorously explore and objectively evaluate all reasonable alternatives, and the reasonable alternative, and for alternatives which were eliminated from detailed study, briefly discuss the reasons for their having been eliminated.

(b) Devote substantial treatment to each alternative considered in detail including the proposed action so that reviewers may evaluate their comparative merits.

(c) Include reasonable alternatives not within the jurisdiction of the lead agency.

(d) Include the alternative of no action.

(e) Identify the agency's preferred alternative or alternatives, if one or more exists, in the draft statement and identify such alternative in the final statement unless another law prohibits the expression of such a preference.

(f) Include appropriate mitigation measures not already included in the proposed action or alternatives.

§ 1502.15 Affected environment.

The environmental impact statement shall succinctly describe the environment of the area(s) to be affected or created by the alternatives under consideration. The descriptions shall be no longer than is necessary to understand the effects of the alternatives. Data and analyses in a statement shall be commensurate with the importance of the impact, with less important material summarized, consolidated, or simply referenced. Agencies shall avoid useless bulk in statements and shall concentrate effort and attention on important issues. Verbose descriptions of the affected environment are themselves no measure of the adequacy of an environmental impact statement.

§ 1502.16 Environmental consequences.

This section forms the scientific and analytic basis for the comparisons under § 1502.14. It shall consolidate the discussions of those elements required by sections 102(2)(C)(i), (ii), (iv), and (v) of NEPA which are within the scope of the statement and as much of section 102(2)(C)(iii) as is necessary to support the comparisons. The discussion will include the environmental impacts of the alternatives including the proposed action, any adverse environmental effects which cannot be avoided should the proposal be implemented, the relationship between short-term uses of man's environment and the maintenance and enhancement of long-term productivity, and any irreversible or irretrievable commitments of resources which would be involved in the proposal should it be implemented. This section should not duplicate discussions in § 1502.14. It shall include discussions of:

(a) Direct effects and their significance (§ 1508.8).

(b) Indirect effects and their significance (§ 1508.8).

(c) Possible conflicts between the proposed action and the objectives of Federal, regional, State and local (and in the case of a reservation, Indian tribe) land use plans, policies and controls for the area concerned. (See § 1506.2(d).)

(d) The environmental effects of alternatives including the proposed action. The comparisons under § 1502.14 will be based on this discussion.

(e) Energy requirements and conservation potential of various alternatives and mitigation measures.

(f) Natural or depletable resource requirements and conservation potential of various alternatives and mitigation measures.

(g) Urban quality, historic and cultural resources, and the design of the built environment, including the reuse and conservation potential of various alternatives and mitigation measures.

(h) Means to mitigate adverse environmental impacts (if not fully covered under § 1502.14(f)).

[43 FR 55994, Nov. 29, 1978; 44 FR 873, Jan. 3, 1979]

§ 1502.17 List of preparers.

The environmental impact statement shall list the names, together with their qualifications (expertise, experience, professional disciplines), of the persons who were primarily responsible for preparing the environmental impact statement or significant background papers, including the basic components of the statement (§§ 1502.6 and 1502.8). Where possible the persons who are responsible for a particular analysis, including analyses in background papers, shall be identified. Normally the list will not exceed two pages.

§ 1502.18 Appendix.

If an agency prepares an appendix to an environmental impact statement the appendix shall:

(a) Consist of material prepared in connection with an environmental impact statement (as distinct from material which is not so prepared and which is incorporated by reference (§ 1502.21)).

(b) Normally consist of material which substantiates any analysis fundamental to the impact statement.

(c) Normally be analytic and relevant to the decision to be made.

(d) Be circulated with the environmental impact statement or be readily available on request.

§ 1502.19 Circulation of the environmental impact statement.

Agencies shall circulate the entire draft and final environmental impact statements except for certain appendices as provided in § 1502.18(d) and unchanged statements as provided in § 1503.4(c). However, if the statement is unusually long, the agency may circulate the summary instead, except that the entire statement shall be furnished to:

(a) Any Federal agency which has jurisdiction by law or special expertise with respect to any environmental impact involved and any appropriate Federal, State, or local agency authorized to develop and enforce environmental standards.

(b) The applicant, if any.

(c) Any person, organization, or agency requesting the entire environmental impact statement.

(d) In the case of a final environmental impact statement any person, organization, or agency which submitted substantive comments on the draft.

If the agency circulates the summary and thereafter receives a timely request for the entire statement and for additional time to comment, the time for that requestor only shall be extended by at least 15 days beyond the minimum period.

§ 1502.20 Tiering.

Agencies are encouraged to tier their environmental impact statements to eliminate repetitive discussions of the same issues and to focus on the actual issues ripe for decision at each level of environmental review (§ 1508.28). Whenever a broad environmental impact statement has been prepared (such as a program or policy statement) and a subsequent statement or environmental assessment is then prepared on an action included within the entire program or policy (such as a site specific action) the subsequent statement or environmental assessment need only summarize the issues discussed in the broader statement and incorporate discussions from the broader statement by reference and shall concentrate on the issues specific to the subsequent action. The subsequent document shall state where the earlier document is available. Tiering may also be appropriate for different stages of actions. (Section 1508.28).

§ 1502.21 Incorporation by reference.

Agencies shall incorporate material into an environmental impact statement by reference when the effect will be to cut down on bulk without impeding agency and public review of the action. The incorporated material shall be cited in the statement and its content briefly described. No material may be incorporated by reference unless it is reasonably available for inspection by potentially interested persons within the time allowed for comment. Material based on proprietary data which is itself not available for review and comment shall not be incorporated by reference.

§ 1502.22 Incomplete or unavailable information.

When an agency is evaluating reasonably foreseeable significant adverse effects on the human environment in an environmental impact statement and there is incomplete or unavailable information, the agency shall always make clear that such information is lacking.

(a) If the incomplete information relevant to reasonably foreseeable significant adverse impacts is essential to a reasoned choice among alternatives and the overall costs of obatining it are not exorbitant, the agency shall include the information in the environmental impact statement.

(b) If the information relevant to reasonably foreseeable significant adverse impacts cannot be obtained because the overall costs of obtaining it are exorbitant or the means to obtain it are now known, the agency shall include within the environmental impact statement:

(1) A statement that such information is incomplete or unavailable; (2) a statement of the relevance of the incomplete or unavailable information to evaluating reasonably foreseeable significant adverse impacts on the human environment; (3) a summary of existing credible scientific evidence which is relevant to evaluating the reasonably foreseeable significant adverse impacts on the human environment, and (4) the agency's evaluation of such impacts based upon theoretical approaches or research methods generally accepted in the scientific community. For the purposes of this section, ''reasonably foreseeable'' includes impacts which have catastrophic consequences, even if their probability of occurrence is low, provided that the analysis of the impacts is supported by credible scientific evidence, is not based on pure conjecture, and is within the rule of reason.

(c) The amended regulation will be applicable to all environmental impact statements for which a Notice of Intent (40 CFR 1508.22) is published in the *Federal Register* on or after May 27, 1986. For environmental impact statements in progress, agencies may choose to comply with the requirements of either the original or amended regulation.

[51 FR 15625, Apr. 25, 1986]

§ 1502.23 Cost-benefit analysis.

If a cost-benefit analysis relevant to the choice among environmentally different alternatives is being considered for the proposed action, it shall be incorporated by reference or appended to the statement as an aid in evaluating the environmental consequences. To assess the adequacy of compliance with section 102(2)(B) of the Act the statement shall, when a cost-benefit analysis is prepared, discuss the relationship between that analysis and any analyses of unquantified environmental impacts, values, and amenities. For purposes of complying with the Act, the weighting of the merits and drawbacks of the varous alternatives need not be displayed in a monetary cost-benefit analysis and should not be when there are important qualitative considerations. In any event, an environmental impact statement should at least indicate those considerations, including factors not related to environmental quality, which are likely to be relevant and important to a decision.

§ 1502.24 Methodology and scientific accuracy.

Agencies shall insure the professional integrity, including scientific integrity, of the discussions and analyses in environmental impact statements. They shall identify any methodologies used and shall make explicit reference by footnote to the scientific and other sources relied upon for conclusions in the statement. An agency may place discussion of methodology in an appendix.

§ 1502.25 Environmental review and consultation requirements.

(a) To the fullest extent possible, agencies shall prepare draft environmental impact statements concurrently with and integrated with environmental impact analyses and related surveys and studies required by the Fish and Wildlife Coordination Act (16 U.S.C. 661 et seq.), the National Historic Preservation Act of 1966 (16 U.S.C. 470 et seq.), the Endangered Species Act of 1973 (16 U.S.C. 1531 et seq.), and other environmental review laws and executive orders.

(b) The draft environmental impact statement shall list all Federal permits, licenses, and other entitlements which must be obtained in implementing the proposal. If it is uncertain whether a Federal permit, license, or other entitlement is necessary, the draft environmental impact statement shall so indicate.

PART 1503 — COMMENTING

Authority: NEPA, the Environmental Quality Improvement Act of 1970, as amended (42 U.S.C. 4371 *et seq.*), sec. 309 of the Clean Air Act, as amended (42 U.S.C. 7609), and Executive Order 11514 (March 5, 1970, as amended by Executive Order 11991, May 24, 1977).

Source: 43 FR 55997, Nov. 29, 1978, unless otherwise noted.

§1503.1 Inviting comments.

(a) After preparing a draft environmental impact statement and before preparing a final environmental impact statement the agency shall:

(1) Obtain the comments of any Federal agency which has jurisdiction by law or special expertise with respect to any environmental impact involved or which is authorized to develop and enforce environmental standards.

(2) Request the comments of:

(i) Appropriate State and local agencies which are authorized to develop and enforce environmental standards;

(ii) Indian tribes, when the effects may be on a reservation; and

(iii) Any agency which has requested that it receive statements on actions of the kind proposed.

Office of Management and Budget Circular A-95 (Revised), through its system of clearinghouses, provides a means of securing the views of State and local environmental agencies. The clearinghouses may be used, by mutal agreement of the lead agency and the clearinghouse, for securing State and local reviews of the draft environmental impact statements.

(3) Request comments from the applicant, if any.

(4) Request comments from the public, affirmatively soliciting comments from those persons or organizations who may be interested or affected.

(b) An agency may request comments on a final environmental impact statement before the decision is finally made. In any case other agencies or persons may make comments before the final decision unless a different time is provided under § 1506.10.

§ 1503.2 Duty to comment.

Federal agencies with jurisdiction by law or special expertise with respect to any environmental impact involved and agencies which are authorized to develop and enforce environment standards shall comment on statements within their jurisdiction, expertise, or authority. Agencies shall comment within the time period specified for comment in § 1506.10. A Federal agency may reply that it has no comment. If a cooperating agency is satisfied that its views are adequately reflected in the environmental impact statement, it should reply that it has no comment.

§ 1503.3 Specificity of comments.

(a) Comments on an environmental impact statement or on a proposed action shall be as specific as possible and may address either the adequacy of the statement or the merits of the alternatives discussed or both.

(b) When a commenting agency criticizes a lead agency's predictive methodology, the commenting agency should describe the alternative methodology which it prefers and why.

(c) A cooperating agency shall specify in its comments whether it needs additional information to fulfill other applicable environmental reviews or consultation requirements and what information it needs. In particular, it shall specify any additional information it needs to comment adequately on the draft statement's analysis of significant site-specific effects associated with the granting or approving by that cooperating agency of necessary Federal permits, licenses, or entitlements.

(d) When a cooperating agency with jurisdiction by law objects to or expresses reservations about the proposal on grounds of environmental impacts, the agency expressing the objection or reservation shall specify the mitigation measures it considers necessary to allow the agency to grant or approve applicable permit, license, or related requirements or concurrences.

§ 1503.4 Response to comments.

(a) An agency preparing a final environmental impact statement shall assess and consider comments both individually and collectively, and shall respond to one or more of the means listed below, stating its reponse in the final statement. Possible response are to:

(1) Modify alternatives including the proposed action.

(2) Develop and evaluate alternatives not previously given serious consideration by the agency.

(3) Supplement, improve, or modify its analyses.

(4) Make factual corrections.

(5) Explain why the comments do not warrant further agency response, citing the sources, authorities, or reasons which support the agency's position and, if appropriate, indicate those circumstances which would trigger agency reappraisal or further response.

(b) All substantive comments received on the draft statement (or summaries thereof where the response has been exceptionally voluminous), should be attached to the final statement whether or not the comment is thought to merit individual discussion by the agency in the text of the statement.

(c) If changes in response to comments are minor and are confined to the responses described in paragraphs (a) (4) and (5) of this section, agencies may write them on errata sheets and attach them to the statement instead of rewriting the draft statement. In such cases only the comments, the responses, the changes and not the final statement need be circulated (§ 1502.19). The entire document with a new cover sheet shall be filed as the final statement (§ 1506.9).

PART 1504 — PREDECISION REFERRALS TO THE COUNCIL OF PROPOSED FEDERAL ACTIONS DETERMINED TO BE ENVIRONMENTALLY UNSATISFACTORY

Sec.

Authority: NEPA, the Environmental Quality Improvement Act of 1970, as amended (42 U.S.C. 4371 et seq.), Sec. 309 of the Clean Air Act, as amended (42 U.S.C. 7609), and Executive order 11514 (March 5, 1970, as amended by Executive Order 11991, May 24, 1977).

Source: 43 FR 55998, Nov. 29, 1978, unless otherwise noted.

§ 1504.1 Purpose.

(a) This part establishes procedures for referring to the Council Federal interagency disagreements concerning proposed major Federal actions that might cause unsatisfactory environmental effects. It provides means for early resolution of such disagreements.

(b) Under section 309 of the Clear Air Act (42 U.S.C. 7609), the Administrator of the Environmental Protection Agency is directed to review and comment publicly on the environmental impacts of Federal activities, including actions for which environmental impact statements are prepared. If after this review the Administrator determines that the matter is "unsatisfactory from the standpoint of public health or welfare or environmental quality," section 309 directs that the matter be referred to the Council (hereafter "environmental referrals").

(c) Under section 102(2)(C) of the Act other Federal agencies may make similar reviews of environmental impact statements, including judgments on the acceptability of anticipated environmental impacts. These reviews must be made available to the President, the Council and the public.

§ 1504.2 Criteria for referral.

Environmental referrals should be made to the Council only after concerted timely (as early as possible in the process), but unsuccessful attempts to resolve differences with the lead agency. In determining what environmental objections to the matter are appropriate to refer to the Council, an agency should weigh potential adverse environmental impacts, considering:

(a) Possible violation of national environmental standards or policies.
(b) Severity.
(c) Geographical scope.
(e) Importance as precedents.
(f) Availability of environmentally preferable alternatives.

§1504.3 Procedure for referrals and response.

(a) A Federal agency making the referral to the Council shall:

(1) Advise the lead agency at the earliest possible time that it intends to refer a matter to the Council unless a satisfactory agreement is reached.

(2) Include such advice in the referring agency's comments on the draft environmental impact statement, except when the statement does not contain adequate information to permit an assessment of the matter's environmental acceptability.

(3) Identify any essential information that is lacking and request that it be made available at the earliest possible time.

(4) Send copies of such advice to the Council.

(b) The referring agency shall deliver its referral to the Council not later than twenty-five (25) days after the final environmental impact statement has been made available to the Environment Protection Agency, commenting agencies, and the public. Except when an extension of this period has been granted by the lead agency, the Council will not accept a referral after that date.

(c) The referral shall consist of:

(1) A copy of the letter signed by the head of the referring agency and delivered to the lead agency informing the lead agency of the referral and the reasons for it, and requesting that no action be taken to implement the matter until the Council acts upon the referral. The letter shall include a copy of the statement reffered to in (c)(2) of this section.

(2) A statement supported by factual evidence leading to the conclusion that the matter is unsatisfactory from the standpoint of public health or welfare or environmental quality.

The statement shall:

(i) Identify any material factors in controversy and incorporate (by reference if appropriate) agreed upon facts,

(ii) Identify any existing environmental requirements or policies which would be violated by the matter,

(iii) Present the reasons why the referring agency belives the matter is environmentally unsatisfactory.

(iv) Contain a finding by the agency whether the issue raised is of national importance because of the threat to national environmental resources or policies or for some other reason,

(v) Review the steps taken by the referring agency to bring its concerns to the attention of the lead agency at the earliest possible time, and

(vi) Give the referring agency's recommendations as to what mitigation alternative, further study, or other course of action (including abandonment of the matter) are necessary to remedy the situation.

(d) Not later than twenty-five (25) days after the referral to the Council the lead agency may deliver a response to the Council and the referring agency. If the lead agency request more time and gives assurance that the matter will not go forward to the interim, the Council may grant an extension. The response shall:

(1) Address fully the issues raised in the referral.

(2) Be supported by evidence.

(3) Give the lead agency's response to the referring agency's recommendations.

(e) Interested persons (including the applicant) may deliver their views in writing to the Council. Views in support of the referral should be delivered not later than the referral. Views in support of the response shall be delivered not later than the response.

(f) Not later than twenty-five (25) days after receipt of both the referral and any response or upon being informed that there will be no response (unless the lead agency agrees to a longer time), the Council may take on or more of the following actions:

(1) Conclude that the process of referral and response has successfully resolved the problem.

(2) Initiate discussions with the agencies with the objective of mediation with referring and lead agencies.

(3) Hold public meetings or hearings to obtain additional views and information.

(4) Determination that the issue is not one of national importance and request the referring and lead agencies to pursue their decision process.

(5) Determine that the issue should be further negotiated by the referring and lead agencies and is not appropriate for Council consideration until one or more heads of agencies report to the Council that the agencies' disagreements are irreconcilable.

(6) Publish its findings and recommendations (including where appropriate a finding that the submitted evidence does not support the position of an agency).

(7) When appropriate, submit the referral and the response together with the Council's recommendation to the President for action.

(g) The Council shall take no longer than 60 days to complete the actions specified in paragraph (f)(2), (3), or (5) of this section.

(h) When the referral involves an action required by statute to be determined on the record after opportunity for agency hearing, the referral shall be conducted in a manner consistent with 5 U.S.C. 557(d) Administrative Procedure Act).

[43 FR 55998, Nov. 29, 1978; 44 FR 873, Jan. 3, 1979]

PART 1505 — NEPA AND AGENCY DECISIONMAKING

Authority: NEPA, the Environmental Quality Improvement Act of 1970, as amended (42 U.S.C. 4371 *et seq.*), sec. 309 of the Clean Air Act, as amended (42 U.S.C. 7609), and Executive Order 11514 (March 5, 1970, as amended by Executive Order 11991, May 24, 1977).

Source: 43 FR 55999, Nov. 29, 1978, unless otherwise noted.

§ 1505.1 Agency decisionmaking procedures.

Agencies shall adopt procedures (§ 1507.3) to ensure that decisions are made in accordance with the policies and purposes of the Act. Such procedures shall include but not be limited to:

(a) Implementing procedures under section 102(2) to achieve the requirements of sections 101 and 102(1).

(b) Designating the major decision points for the agency's principal programs likely to have a significant effect on the human environment and assuring that the NEPA process corresponds with them.

(c) Requiring that relevant environmental documents, comments, and responses be part of the record in formal rulemaking or adjudicatory proceedings.

(d) Requiring that relevant environmental documents, comments, and responses accompany the proposal through existing agency review processes so that agency officials use the statement in making decisions.

(e) Requiring that the alternatives considered by the decisionmaker are encompassed by the range of alternatives discussed in the relevant environmental documents and that the decisionmaker consider the alternatives described in the environmental impact statement. If another decision document accompanies the relevant environmental documents to the decisionmaker, agencies are encouraged to make available to the public before the decision is made any part of that document that relates to the comparison of alternatives.

§ 1505.2 Record of decision in cases requiring environmental impact statements.

At the time of its decision (§ 1506.10) or, if appropriate, its recommendation to Congress, each agency shall prepare a concise public record of decision. The record, which may be integrated into any other record prepared by the agency, including that required by OMB Circular A-95 (Revised), part I, sections 6(c) and (d), and part II, section 5(b)(4), shall:

(a) State what the decision was.

(b) Identify all alternatives considered by the agency in reaching its decision, specifying the alternative or alternatives which were considered to be environmentally preferable. An agency may discuss preferences among alternatives based on relevant factors including economic and technical considerations and agency statutory missions. An agency shall identify and discuss all such factors including any essential considerations of national policy which were balanced by the agency in making its decision and state how those considerations entered into its decision.

(c) State whether all practicable means to avoid or minimize environmental harm from the alternative selected have been adopted, and if not, why they were not. A monitoring and enforcement program shall be adopted and summarized where applicable for any mitigation.

§ 1505.3 Implementing the decision.

Agencies may provide for monitoring to assure that their decisions are carried out and should do so in important cases. Mitigation (§ 1505.2(c)) and other conditions established in the environmental impact statement or during its review and committed as part of the decision shall be implemented by the lead agency or other appropriate consenting agency. The lead agency shall:

(a) Include appropriate conditions in grants, permits or other approvals.

(b) Condition funding of actions on mitigation.

(c) Upon request, inform cooperating or commenting agencies on progress in carrying out mitigation measures which they have proposed and which were adopted by the agency making the decision.

(d) Upon request, make available to the public the results of relevant monitoring.

PART 1506 — OTHER REQUIREMENTS OF NEPA

Sec.

Authority: NEPA, the Environmental Quality Improvement Act of 1970, as amended (42 U.S.C. 4371 et seq.), Section 309 of the Clean Air Act, as amended (42 U.S.C. 7609), and Executive Order 11514, Protection and Enhancement of Environmental Quality (March 5, 1970, as amended by Executive Order 11991, May 24, 1977).

Source: 43 FR 56000, Nov. 29, 1978, unless otherwise noted.

§ 1506.1 Limitations on actions during NEPA process.

(a) Until an agency issues a record of decision as provided in § 1505.2 (except as provided in paragraph (c) of this section), no action concerning the proposal shall be taken which would:

(1) Have an adverse environmental impact; or

(2) Limit the choice of reasonable alternatives.

(b) If any agency is considering an application from a non-Federal entity, and is aware that an action within the agency's jurisdiction that would meet either of the criteria in paragraph (a) of this section, then the agency shall promptly notify the applicant that the agency will take appropriate action to insure that the objectives and procedures of NEPA are achieved.

(c) While work on a required program environmental impact statement is in progress and the action is not covered by an existing program statement, agencies shall not undertake in the interim any major Federal action covered by the program which may significantly affect the quality of the human environment unless such action:

(1) Is justified independently of the program;

(2) Is itself accompanied by an adequate environmental impact statement; and

(3) Will not prejudice the ultimate decision on the program. Interim action prejudices the ultimate decision on the program when it tends to determine subsequent development or limit alternatives.

(d) This section does not preclude development by applicants of plans or designs or performance of other work necessary to support an application for Federal, State or local permits or assistance. Nothing in this section shall preclude Rural Electrification Administration approval of minimal expenditures not affecting the environment (e.g., long leadtime equipment and purchase options) made by non-governmental entities seeking loan guarantees from the Administration.

§ 1506.2 Elimination of duplication with State and local procedures.

(a) Agencies authorized by law to cooperate with State agencies of statewide jurisdiction pursuant to section 102(2)(d) of the Act may do so.

(b) Agencies shall cooperate with State and local agencies to the fullest extent possible to reduce duplication between NEPA and State and local requirements, unless the agencies are specifically barred from doing so by some other law. Except for cases covered by paragraph (a) of this section, such cooperation shall to the fullest extent possible include:

(1) Joint planning processes.

(2) Joint environmental research and studies.

(3) Joint public hearings (except where otherwise provided by statute).

(4) Joint environmental assessments.

(c) Agencies shall cooperate with State and local agencies to the fullest extent possible to reduce duplication between NEPA and comparable State and local requirements, unless the agencies are specifically barred from doing so by some other law. Except for cases covered by paragraph (a) of this section, such cooperation shall to the fullest extent possible include joint environmental impact statements. In such cases one or more Federal agencies and one or more State or local agencies shall be joint lead agencies. Where State laws or local ordinances have environmental impact statement requirements in addition to but not in conflict with those in NEPA, Federal agencies shall cooperate in fulfilling these requirements as well as those of Federal laws so that one document will comply with all applicable laws.

(d) To better integrate environmental impact statements into State or local planning processes, statements shall discuss any inconsistency of a proposed action with any approved State or local plan and laws (whether or not federally sanctioned). Where an inconsistency exists, the statement should describe the extent to which the agency would reconcile its proposed action with the plan or law.

§ 1506.3 Adoption.

(a) An agency may adopt a Federal draft of final environmental impact statement or portion thereof provided that the statement or portion thereof meets the standards for an adequate statement under these regulations.

(b) If the actions covered by the original environmental impact statement and the proposed action are substantially the same, the agency adopting another agency's statement is not required to recirculate it except as a final statement. Otherwise the adopting agency shall treat the statement as a draft and recirculate it (except as provided in paragraph (c) of this section).

(c) A cooperating agency may adopt without recirculating the environmental impact statement of a lead agency when, after an independent review of the statement, the cooperating agency concludes that its comments and suggestions have been satisfied.

(d) When an agency adopts a statement which is not final within the agency that prepared it, or when the action it assesses is the subject of referral under Part 1504, or when the statement's adequacy is the subject of a judicial action which is not final, the agency shall so specify.

§1506.4 Combining documents.

Any environmental document in compliance with NEPA may be combined with any other agency document to reduce duplication and paperwork.

§ 1506.5 Agency responsibility.

(a) *Information*. If an agency requires an applicant to submit environmental information for possible use by the agency in preparing an environmental impact statement, then the agency should assist the applicant by outlining the types of information required. The agency shall independently evaluate the information submitted and shall be responsible for its accuracy. If the agency chooses to use the information submitted by the applicant in the environmental impact statement, either directly or by reference, then the names of the persons responsible for the independent evaluation shall be included in the list of preparers (§ 1502.17). It is the intent of this subparagraph that acceptable work not be redone, but that it be verified by the agency.

(b) *Environmental assessments*. If an agency permits an applicant to prepare an environmental assessment, the agency, besides fulfilling the requirements of paragraph (a) of this section, shall make its own evaluation of the environmental issues and take responsibility for the scope and content of the environmental assessment.

(c) *Environmental impact statements*. Except as provided in §§ 1506.2 and 1506.3 any environmental impact statement prepared pursuant to the requirements of NEPA shall be prepared directly by or by a contractor selected by the lead agency or where appropriate under § 1501.6(b), a cooperating agency. It is the intent of these regulations that the contractor be chosen solely by the lead agency, or by the lead agency in cooperation with cooperating agencies, or where appropriate by a cooperating agency to avoid any conflict of interest. Contractors shall execute a disclosure statement prepared by the lead agency, or where appropriate the cooperating agency, specifying that they have no financial or other interest in the outcome of the project. If the document is prepared by contract, the responsible Federal official shall furnish guidance and participate in the preparation and shall independently evaluate the statement prior to its approval and take responsibility for its scope and contents. Nothing in this section is intended to prohibit any agency from requesting any person to submit information to it or to prohibit any person from submitting information to any agency.

§ 1506.6 Public involvement.

Agencies shall:

(a) Make diligent efforts to involve the public in preparing and implementing their NEPA procedures.

(b) Provide public notice of NEPA-related hearings, public meetings, and the availability of environmental documents so as to inform those persons and agencies who may be interested or affected.

(1) In all cases the agency shall mail notice to those who have requested it on an individual action.

(2) In the case of an action with effects of national concern notice shall include publication in the *Federal Register* and notice by mail to national organizations reasonably expected to be interested in the matter and may include listing in the *102 Monitor*. An agency engaged in rulemaking may provide notice by mail to national organizations who have requested that notice regularly be provided. Agencies shall maintain a list of such organizations.

(3) In the case of an action with effects primarily of local concern the notice may include:

(i) Notice to State and areawide clearinghouses pursuant to OMB Circular A-95 (revised).

(ii) Notice to Indian tribes when effects may occur on reservations.

(iii) Following the affected State's public notice procedures for comparable actions.

(iv) Publication in local newspapers (in papers of general circulation rather than legal papers).

(v) Notice through other local media.

(vi) Notice to potentially interested community organizations including small business associations.

(vii) Publication in newsletters that may be expected to reach potentially interested persons.

(viii) Direct mailing to owners and occupants to nearby or affected property.

(ix) Posting of notice on and off site in the area where the action is to be located.

(c) Hold or sponsor public hearings or public meetings whenever appropriate or in accordance with statutory requirements applicable to the agency. Criteria shall include whether there is:

(1) Substantial environmental controversy concerning the proposed action or substantial interest in holding the hearing.

(2) A request for a hearing by another agency with jurisdiction over the action supported by reasons why a hearing will be helpful. If a draft environmental impact statement is to be considered at a public hearing, the agency should make the statement available to the public at least 15 days in advance (unless the purpose of the hearing is to provide information for the draft environmental impact statement).

(d) Solicit appropriate information from the public.

(e) Explain in its procedures where interested persons can get information or status reports on environmental impact statements and other elements of the NEPA process.

(f) Make environmental impact statements, the comments received, and any underlying documents to the provisions of the Freedom of Information Act (5 U.S.C. 552), without regard to the exclusion for interagency memoranda where such memoranda transmit comments of Federal agencies on the environmental impact of the proposed action. Materials to be made available to the public shall be provided to the public without charge to the extent practicable, or at a fee which is not more than the actual costs of reproducing copies required to be sent to other Federal agencies, including the Council.

§ 1506.7 Further guidance.

The Council may provide further guidance concerning NEPA and its procedures including:

(a) A handbook which the Council may supplement from time to time, which shall in plain language provide guidance and instructions concerning the application of NEPA and these regulations.

(b) Publication of the Council's Memoranda to Heads of Agencies.

(c) In conjunction with the Environmental Protection Agency and the publication of the 102 Monitor, notice of:

(1) Research activities;

(2) Meetings and conferences related to NEPA; and

(3) Successful and innovative procedures used by agencies to implement NEPA.

§ 1506.8 Proposals for legislation.

(a) The NEPA process for proposals for legislation (§ 1508.17) significantly affecting the quality of the human environment shall be integrated with the legislative process of the Congress. A legislative environmental impact statement is the detailed statement required by law to be included in a recommendation or report on a legislative proposal to Congress. A legislative environmental impact statement shall be considered part of the formal transmittal of a legislative proposal to Congress; however, it may be transmitted to Congress up to 30 days later in order to allow time for completion of an accurate statement which can serve as the basis for public and Congressional debate. The statement must be available in time for Congressional hearings and deliberations.

(b) Preparation of a legislative environmental impact statement shall conform to the requirements of these regulations except as follows:

(1) There need not be a scoping process.

(2) The legislative statement shall be prepared in the same manner as a draft statement, but shall be considered the "detailed statement" required by statute; *Provided*, That when

any of the following conditions exist both the draft and final environmental impact statement on the legislative proposal shall be prepared and circulated as provided by §§ 1503.1 and 1506.10.

(i) A Congressional Committee with jurisdiction over the proposal has a rule requiring both draft and final environmental impact statements.

(ii) The proposal results from a study process required by statute (such as those required by the Wild and Scenic Rivers Act (16 U.S.C. 1271 et seq.) and the Wilderness Act (16 U.S.C. 1131 et seq.)).

(iii) Legislative approval is sought for Federal or federally assisted construction or other projects which the agency recommends be located at specific geographic locations. For proposals requiring an environmental impact statement for the acquisition of space by the General Services Administration, a draft statement shall accompany the Prospectus or the 11(b) Report of Building Project Surveys to the Congress, and a final statement shall be completed before site acquisition.

(iv) The agency decides to prepare draft and final statements.

(c) Comments on the legislative statement shall be given to the lead agency which shall forward them along with its own responses to the Congressional committees with jurisdiction.

§ 1506.9 Filing requirements.

Environmental impact statements together with comments and responses shall be filed with the Environmental Protection Agency, attention Office of Federal Activities (A-104), 401 M Street, S.W., Washington, D.C. 20460. Statements shall be filed with EPA no earlier than they are also transmitted to commenting agenices and made available to the public. EPA shall deliver one copy of each statement to the council, which shall satisfy the requirement of availability to the President. EPA may issue guidelines to agencies to implement its responsibilities under this section and § 1506.10.

§ 1506.10 Timing of agency action.

(a) The Environmental Protection Agency shall publish a notice in the *Federal Register* each week of the environmental impact statements filed during the preceding week. The minimum time periods set forth in this section shall be calculated from the date of publication of this notice.

(b) No decision on the proposed action shall be made or recorded under § 1505.2 by a Federal agency until the latter of the following dates:

(1) Ninety (90) days after publication of the notice described above in paragraph (a) of this section for a draft environmental impact statement.

(2) Thirty (30) days after publication of the notice described above in paragraph (a) of this section for final environmental impact statement.

An exception to the rules on timing may be made in the case of an agency decision which is subject to a formal internal appeal. Some agencies have a formally established appeal process which allows other agencies or the public to take appeals on a decision and make their views known, after publication of the final environmental impact statement. In such cases, where a real opportunity exists to alter the decision, the decision may be made and recorded at the same time the environmental impact statement is published. This means that the period of appeal of the decision and the 30-day period prescribed in paragraph (b)(2) of this section may run concurrently. In such cases the environmental impact statement shall explain the timing and the public's right of appeal. An agency engaged in rulemaking under the Administrative Procedure Act or other statute for the purpose of protecting the public health or safety, may waive the time period in paragraph (b)(2) of this section and publish a decision on the final rule simultaneously with publication of the notice of the availability of the final environmental impact statement as described in paragraph (a) of this section.

(c) If the final environmental impact statement is filed within ninety (90) days after a draft environmental impact statement is filed with the Environmental Protection Agency, the minimum thirty (30) day period and the minimum ninety (90) day period may run concurrently. However, subject to paragraph (d) of this section agencies shall allow not less than 45 days for comments on draft statements.

(d) The lead agency may extend prescribed periods. The Environmental Protection Agency may upon a showing by the lead agency of compelling reasons of national policy reduce the prescribed periods and may upon a showing by any other Federal agency of compelling reasons of national policy also extend prescribed periods, but only after consultation with the lead agency. (Also see § 1507.3(d).) Failure to file timely comments shall not be a sufficient reason for extending a period. If the lead agency does not concur with the extention of time, EPA may not extend it for more than 30 days. When the Environmental Protection Agency reduces or extends any period of time it shall notify the Council.

[43 FR 56000, No. 29, 1978; 44 FR 874, Jan. 3, 1979]

§ 1506.11 Emergenices.

Where emergency circumstances make it necessary to take an action with significant environmental impact without observing the provisions of these regulations, the Federal agency taking the action should consult with the Council about alternative arrangements. Agencies and the Council will limit such arrangements to actions necessary to control the immediate impacts of the emergency. Other actions remain subject to NEPA review.

§ 1506.12 Effective date.

The effective date of these regulations is July 30, 1979, except that for agencies that administer programs that qualify under section 102(2)(D) of the Act or under sec. 104(h) of the Housing and Community Development Act of 1974 an additional four months shall be allowed for the State or local agencies to adopt their implementing procedures.

(a) These regulations shall apply to the fullest extent practicable to ongoing activities and environmental documents begun before the effective date. These regulations do not apply to an environmental impact statement or supplement if the draft statement was filed before the effective date of these regulations. No completed environmental documents need be redone by reason of these regulations. Until these regulations are applicable, the Council's guidelines published in the *Federal Register* of August 1, 1973, shall continue to be applicable. In cases where these regulations are applicable the guidelines are superseded. However, nothing shall prevent an agency from proceeding under these regulations at an earlier time.

(b) NEPA shall continue to be applicable to actions begun before January 1, 1970, to the fullest extent possible.

PART 1507 — AGENCY COMPLIANCE

Authority: NEPA, the Environmental Quality Improvement Act of 1970, as amended (42 U.S.C. 4371 *et seq.*), sec. 309 of the Clean Air Act, as amended (42 U.S.C. 7609), and Executive Order 11514 (March 5, 1970, as amended by Executive Order 11991, May 24, 1977).

Source: 43 FR 56002, Nov. 29, 1978, unless otherwise noted.

§ 1507.1 Compliance.

All agencies of the Federal Government shall comply with these regulations. It is the intent of these regulations to allow each agency flexibility in adapting its implementing procedures authorized by § 1507.3 to the requirements of other applicable laws.

§ 1507.2 Agency capability to comply.

Each agency shall be capable (in terms of personnel and other resources) of complying with the requirements enumerated below. Such compliance may include use of other's resources, but the using agency shall itself have sufficient capability to evaluate what others do for it. Agencies shall:

(a) Fulfill the requirements of section 102(2)(A) of the Act to utilize a systematic, interdisciplinary approach which will insure the integrated use of the natural and social sciences and the environmental design arts in planning and in decisionmaking which may have an impact on the human environment. Agencies shall designate a person to be responsible for overall review of agency NEPA compliance.

(b) Identify methods and procedures required by section 102(2)(B) to insure that presently unquantified environmental amenities and values may be given appropriate consideration.

(c) Prepare adequate environmental impact statements pursuant to section 102(2)(C) and comment on statements in the areas where the agency has jurisdiction by law or special expertise or is authorized to develop and enforce environmental standards.

(d) Study, develop, and describe alternatives to recommended courses of action in any proposal which involves unresolved conflicts concerning alternative uses of available resources. This requirement of section 102(2)(E) extends to all such proposals, not just the more limited scope of section 102(2)(C)(iii) where the discussion of alternatives is confined to impact statements.

(e) Comply with the requirements of section 102(2)(H) that the agency initiate and utilize ecological information in the planning and development of resource-oriented projects.

(f) Fulfill the requirements of sections 102(2)(F), 102(2)(G), and 102(2)(I), of the Act and of Executive Order 11514, Protection and Enhancement of Environmental Quality, Sec. 2.

§ 1507.3 Agency procedures.

(a) Not later than eight months after publication of these regulations as finally adopted in the *Federal Register*, or five months after the establishment of an agency, whichever shall come later, each agency shall as necessary adopt procedures to supplement these regulations. When the agency is a department, major subunits are encouraged (with the consent of the department) to adopt their own procedures. Such procedures shall not paraphrase these regulations. They shall confine themselves to implementing procedures. Each agency shall consult with the Council while developing its procedures and before publishing them in the *Federal Register* for comment. Agencies with similar programs should consult with each other and the Council to coordinate their procedures, especially for programs requesting similar information from applicants. The procedures shall be adopted only after an opportunity for public review and after review by the Council for conformity with the Act and these regulations. The Council shall complete its review within 30 days. Once in effect they shall be filed with the Council and made readily available to the public. Agencies are encouraged to publish explanatory guidance for these regulations and their own procedures. Agencies shall continue to review their policies and procedures and in consultation with the Council to revise them as necessary to ensure full compliance with the purposes and provisions of the Act.

(b) Agency procedures shall comply with these regulations except where compliance would be inconsistent with statutory requirements and shall include:

(1) Those procedures required by §§ 1501.2(d), 1502.9(c)(3), 1501.1, 1506.6(e), and 1508.4.

(2) Specific criteria for and identification of those typical classes of action:

(i) Which normally do require environmental impact statements.

(ii) Which normally do not require either an environmental impact statement or an environmental assessment (categorical exclusions (§ 1508.4)).

(iii) Which normally require environmental assessments but not necessarily environmental impact statements.

(c) Agency procedures may include specific criteria for providing limited exceptions to the provisions of these regulations for classified proposals. They are proposed actions which are specifically authorized under criteria established by an Executive Order or statue to be kept secret in the interest of national defense or foreign policy and are in fact properly classified pursuant to such Executive Order or statute. Environmental assessments and environmental impact statements which address classified proposals may be safeguarded and restricted from public dissemination in accordance with agencies' own regulations applicable to classified information. These documents may be organized so that classified portions can be included as annexes, in order that the unclassified portions can be made available to the public.

(d) Agency procedures may provide for periods of time other than those presented in § 1506.10 when necessary to comply with other specific statutory requirements.

(e) Agency procedures may provide that where there is a lengthy period between the agency's decision to prepare an environmental impact statement and the time of actual preparation, the notice of intent required by § 1501.7 may be published at a reasonable time in advance of preparation of the draft statement.

PART 1508 — TERMINOLOGY AND INDEX

Sec.

Authority: NEPA, the Environmental Quality Improvement Act of 1970, as amended (42 U.S.C. 4371 *et seq.*), sec. 309 of the Clean Air Act, as amended (42 U.S.C. 7609), and Executive Order 11514 (March 5, 1970, as amended by Executive Order 1191, May 24, 1977).

Source: 43 FR 56003, Nov. 29, 1978, unless otherwise noted.

§ 1508.1 Terminology.

The terminology of this part shall be uniform throughout the Federal Government.

§ 1508.2 Act.

''Act'' means the National Environmental Policy Act, as amended (42 U.S.C. 4321, et seq.) which is also referred to as ''NEPA.''

§ 1508.3 Affecting.

''Affecting'' means will or may have an effect on.

§ 1508.4 Categorical exclusion.

''Categorical Exclusion'' means a category of actions which do not individually or cumulatively have a significant effect on the human environment and which have been found to have no such effect in procedures adopted by a Federal agency in implementation of these regulations (§ 1507.3) and for which, therefore, neither an environmental assessment nor an environmental impact statement is required. An agency may decide in its procedures or otherwise, to prepare environmental assessments for the reasons stated in § 1508.9 even though it is not required to do so. Any procedures under this section shall provide for extraordinary circumstances in which a normally excluded action may have a significant environmental effect.

§ 1508.5 Cooperating agency.

''Cooperating Agency'' means any Federal agency other than a lead agency which has jurisdiction by law or special expertise with respect to any environmental impact involved in a proposal (or a reasonable alternative) for legislation or other major Federal action significantly affecting the quality of the human environment. The selection and responsibilities of a cooperating agency are described in § 1501.6. A State or local agency of similar qualifications or, when the effects are on a reservations, an Indian Tribe, may by agreement with the lead agency become a cooperating agency.

§ 1508.6 Council.

''Council'' means the Council on Environmental Quality established by Title II of the Act.

§ 1508.7 Cumulative impact.

''Cumulative impact'' is the impact on the environment which results from the incremental impact of the action when added to other past, present, and reasonably foreseeable future actions regardless of what agency (Federal or non-Federal) or person undertakes such other actions. Cumulative impacts can results from individually minor but collectively significant actions taking place over a period to time.

§ 1508.8 Effects.

''Effects'' include:

(a) Direct effects, which are caused by the action and occur at the same time and place.

(b) Indirect effects, which are caused by the action and are later in time or farther removed in distance, but are still reasonably foreseeable. Indirect effects may include growth inducing effects and other effects related to induced changes in the pattern of land use, population density or growth rate, and related effects on air and water and other natural systems, including ecosystems.

Effects and impacts as used in these regulations are synonymous. Effects includes ecological (such as the effects on natural resources and on the components, structure, and functioning of affected ecosystems), aesthetic, historic, cultural, economic, social, or health, whether direct, direct, or cumulative. Effects may also include those resulting from actions which may have both beneficial and detrimental effects, even if on balance the agency believes that the effect will be beneficial.

§ 1508.9 Environmental assessment.

''Environmental Assessment'':

(a) Means a concise public document for which a Federal agency is responsible that serves to:

(1) Briefly provide sufficient evidence and analysis for determining whether to prepare an environmental impact statement or a finding of no significant impact.

(2) Aid an agency's compliance with the act when no environmental impact statement is necessary.

(3) Facilitate preparation of a statement when one is necessary.

(b) Shall include brief discussions of the need for the proposal, of alternatives as required by section 102(2)(E), of the environmental impacts of the proposed action and alternatives, and a listing of agencies and persons consulted.

§ 1508.10 Environmental document.

''Environmental document'' includes the documents specified in § 1508.9 (environmental assessment), § 1508.11 (environmental impact statement), § 1508.13 (finding of no significant impact), and § 1508.22 (notice of intent).

§ 1508.11 Environmental impact statement.

''Environmental Impact Statement'' means a detailed written statement as required by section 102(2)(C) of the Act.

§ 1508.12 Federal agency.

''Federal agency'' means all agencies of the Federal Government. It does not mean the Congress, the Judiciary, or the President, including the performance of staff functions for the President in his Executive Office. It also includes for purposes of these regulations States and units of general local government and Indian tribes assuming NEPA responsibilities under section 104(h) of the Housing and Community Development Act of 1974.

§ 1508.13 Finding of no significant impact.

''Finding of no significant impact'' means a document by a Federal agency briefly presenting the reasons why an action, not otherwise excluded (§ 1508.4), will not have a significant effect on the human environment and for which an environmental impact statement therefore will not be prepared. It shall include the environmental assessment or a summary of it and shall note any other environmental documents related to it (§ 1501.7(a)(5)). If the assessment is included, the finding need not repeat any of the discussion in the assessment but may incorporate it by reference.

§ 1508.14 Human environment.

''Human Environment'' shall be interpreted comprehensibly to include the natural and physical environment and the relationship of people with that environment. (See the definition of ''effects'' (§ 1508.8)). This means that economic or social effects are not intended by themselves to require preparation of an environmental impact statement. When an environmental impact statement is prepared and economic or social and natural or physical environmental effects are interrelated, then the environmental impact statement will discuss all of these effects on the human environment.

§ 1508.15 Jurisdiction by law.

''Jurisdiction by law'' means agency authority to approve, veto, or finance all or part of the proposal.

§ 1508.16 Lead agency.

''Lead agency'' means the agency or agencies preparing or having taken primary responsibility for preparing the environmental impact statement.

§ 1508.17 Legislation.

''Legislation'' includes a bill or legislative proposal to congress developed by or with the significant cooperation and support of a Federal agency, but does not include request for appropriations. The test for significant cooperation is whether the proposal is in fact predominantly that of the agency rather than another source. Drafting does not by itself constitute significant cooperation. Proposals for legislation include request for ratification of treaties. Only the agency which has primary responsibility for the subject matter involved will prepare a legislative environmental impact statement.

§ 1508.18 Major Federal action.

''Major Federal action'' includes actions with effects that may be major and which are potentially subject to Federal control and responsibility. Major reinforces but does not have a meaning independent of significantly (§ 1508.27). Actions include the circumstance where the responsible officials fail to act and that failure to act is reviewable by courts or administrative tribunals under the Administrative Procedure Act or other applicable law as agency action.

(a) Actions include new and continuing activities, including project and program entirely or partly financed, assisted, conducted, regulated, or approved by federal agencies; new or revised agency rules, regulations, plans, policies, or procedures; and legislative proposals (§ 1506.8, 1508.17). Actions do not include funding assistance solely in the form of general revenue sharing funds, distributed under the State and Local Fiscal Assistance Act of 1972, 31 U.S.C. 1221 et seq., with no Federal agency control over the subsequent use of such funds. Actions do not include bringing judicial or administrative civil or criminal enforcement actions.

(b) Federal actions tend to fall within one of the following categories:

(1) Adoption of official policy, such as rules, regulations, and interpretations adopted pursuant to the Administrative Procedure Act, 5 U.S.C. 551 et seq.; treaties and international conventions or agreements; formal documents establishing an agency's policies which will result in or substantially alter agency programs.

(2) Adoption of formal plans, such as official documents prepared or approved by federal agencies which guide or prescribe alternative uses of federal resources, upon which future agency actions will be based.

(3) Adoption of programs, such as a group of concerted actions to implement a specific policy or plan; systematic and connected agency decisions allocating agency resources to implement a specific statutory program or executive directive.

(4) Approval of specific projects, such as construction or management activities located in a defined geographic area. Projects include actions approved by permit or other regulatory decision as well as federal and federally assisted activities.

§ 1508.19 Matter.

"Matter" includes for purposes of Part 1504:

(a) With respect to the Environmental Protection Agency, any proposed legislation, project, action or regulation as those terms are used in section 309(a) of the Clean Air Act (42 U.S.C. 7609).

(b) With respect to all other agencies, any proposed major federal action to which section 102(2)(C) of NEPA applies.

§ 1508.20 Mitigation.

"Mitigation" includes:

(a) Avoiding the impact altogether by not taking a certain action or parts of an action.

(b) Minimizing impacts by limiting the degree of magnitude of the action and its implementation.

(c) Rectifying the impact by repairing, rehabilitating, or restoring the affected environment.

(d) Reducing or eliminating the impact over time by preservation and maintenance operations during the life of the action.

(e) Compensating for the impact by replacing or providing substitute resources or environments.

§ 1508.21 NEPA process.

"NEPA process" means all measures necessary for compliance with the requirements of Section 2 and Title I of NEPA.

§ 1508.22 Notice of intent.

"Notice of Intent" means a notice that an environmental impact statement will be prepared and considered. The notice shall briefly:

(a) Describe the proposed action and possible alternatives.

(b) Describe the agency's proposed scoping process including whether, when, and where any scoping meeting will be held.

(c) State the name and address of a person within the agency who can answer questions about the proposed action and the environmental impact statement.

§ 1508.23 Proposal.

"Proposal" exists at that stage in the development of an action when an agency subject to the Act has a goal and is actively preparing to make a decision on one or more alternative means of accomplishing that goal and the effects can be meaningfully evaluated. Preparation of an environmental impact statement on a proposal should be timed (§ 1502.5) so that the final statement may be completed in time for the statement to be included in any recommendation or report on the proposal. A proposal may exist in fact as well as by agency declaration that one exists.

§ 1508.24 Referring agency.

"Referring agency" means the federal agency which has referred any matter to the Council after a determination that the matter is unsatisfactory from the standpoint of public health or welfare or environmental quality.

§ 1508.25 Scope.

Scope consists of the range of actions, alternatives, and impacts to be considered in an environmental impact statement. The scope of an individual statement may depend on its relationships to other statements (§§ 1508.20 and 1508.28). To determine the scope of environmental impact statements, agencies shall consider 3 types of actions, 3 types of alternatives, and 3 types of impacts. They include:

(a) Actions (other than unconnected single actions) which may be:

(1) Connected actions, which means that they are closely related and therefore should be discussed in the same impact statement. Actions are connected if they:

(i) Automatically trigger other actions which may require environmental impact statements.

(ii) Cannot or will not proceed unless other actions are taken previously or simultaneously.

(iii) Are interdependent parts of a larger action and depend on the larger action for their justification.

(2) Cumulative actions, which when viewed with other proposed actions have cumulatively significantly impacts and should therefore be discussed in the same impact statement.

(3) Similar actions, which when viewed with other reasonably foreseeable or proposed agency actions, have similarities that provide a basis for evaluating their environmental consequences together, such as common timing or geography. An agency may wish to analyze these actions in the same impact statement. It should do so when the best way to assess adequately the combined impacts of similar actions or reasonable alternatives to such actions is to treat them in a single impact statement.

(b) Alternatives, which include: (1) No action alternative.

(2) Other reasonable courses of actions.

(3) Mitigation measures (not in the proposed action).

(c) Impacts, which may be: (1) Direct; (2) indirect; (3) cumulative.

§ 1508.26 Special expertise.

''Special expertise'' means statutory responsibility, agency mission, or related program experience.

§ 1508.27 Significantly.

''Significantly'' as used in NEPA requires considerations of both context and intensity:

(a) *Context*. This means that the significance of an action must be analyzed in several contexts such as society as a whole (human, national), the affected region, the affected interests, and the locality. Significance varies with the setting of the proposed action. For instance, in the case of a site-specific action, significance would usually depend upon the effects in the locale rather than in the world as a whole. Both short- and long-term effects are relevant.

(b) *Intensity*. This refers to the severity of impact. Responsible officials must bear in mind that more than one agency may make decisions about partial aspects of a major action. The following should be considered in evaluating intensity:

(1) Impacts that may be both beneficial and adverse. A significant effect may exist even if the Federal agency believes that on balance the effect will be beneficial.

(2) The degree to which the proposed action affects public health or safety.

(3) Unique characteristics of the geographic area such as proximity to historic or cultural resources, park lands, prime farmlands, wetlands, wild and scenic rivers, or ecologically critical areas.

(4) The degree to which the effects on the quality of the human environment are likely to be highly controversial.

(5) The degree to which the possible effects on the human environment are highly uncertain or involved unique or unknown risks.

(6) The degree to which the action may establish a precedent for future actions with significant effects or represents a decision in principle about a future consideration.

(7) Whether the action is related to other actions with individually insignificant but cumulatively significant impact. Significance exists if it is reasonable to anticipate a cumulatively significant impact on the environment. Significance cannot be avoided by terming an action temporary or by breaking it down into small component parts.

(8) The degree to which the action may adversely affect districts, sites, highways, structures, or objects listed in or eligible for listing in National Register of Historic Place or may cause loss or destruction of significant scientific, cultural, or historical resources.

(9) The degree to which the action may adversely affect an endangered or threatened species or its habitat that has been determined to be critical under the Endangered Species Act of 1973.

(10) Whether the action threatens a violation of Federal, State, or local law or requirements imposed for the protection of the environment.

§ 1508.28 Tiering.

"Tiering" refers to the coverage of general matters in broader environmental impact statements (such as national program or policy statements) with subsequent narrower statements or environmental analyses (such as regional or basinwide program statements or ultimately site-specific statements) incorporating by reference the general discussions and concentrating solely on the issue specific to the statement subsequently prepared. Tiering is appropriate when the sequence of statements or analyses is:

(a) From a program, plan or policy environmental impact statement to a program, plan, or policy statement or analysis of lesser scope or to a site-specific statement or analysis.

(b) From an environmental impact statement on a specific action at an early stage (such as need and site selection) to a supplement (which is preferred) or a subsequent statement or analysis at a later stage (such as environmental mitigation). Tiering in such cases is appropriate when it helps the lead agency to focus on the issues which are ripe for decision and exclude from consideration issues already decided or not yet ripe.

INDEX TO CEQ REGULATIONS

THE ENVIRONMENTAL QUALITY IMPROVEMENT ACT OF 1970*

TITLE II — ENVIRONMENTAL QUALITY (OF THE WATER QUALITY IMPROVEMENT ACT OF 1974)

Short Title

Sec. 201. This title may be cited as the ''Environmental Quality Improvement Act of 1970.''

Findings, Declarations, and Purposes

Sec. 202. (a) The Congress finds —

(1) That man has caused changes in the environment;

(2) That many of these changes may affect the relationship between man and his environment; and

(3) That population increases and urban concentration contribute directly to pollution and the degradation of our environment.

(b) (1) The Congress declares that there is a national policy for the environment which provides for the enhancement of environmental quality. This policy is evidenced by statutes heretofore enacted relating to the prevention, abatement, and control of environmental pollution, water and land resources, transportation, and economic and regional development.

(2) The primary responsibility for implementing this policy rests with State and local governments.

(3) The Federal Government encourages and supports implementation of this policy through appropriate regional organizations established under existing law.

(c) The purposes of this title are —

(1) To assure that each Federal department and agency conducting or supporting public works activities which affect the environment shall implement the policies established under existing law; and

(2) To authorize an Office of Environmental Quality, which, notwithstanding any other provision of law, shall provide the professional and administrative staff for the Council on Environmental Quality established by Public Law 91-190.

Office of Environmental Quality

Sec. 203. (a) There is established in the Executive Office of the President an office to be known as the Office of Environmental Quality (hereafter in this title referred to as the ''Office''). The Chairman of the Council on Environmental Quality established by Public Law 91-190 shall be the Director of the Office. There shall be in the Office a Deputy Director who shall be appointed by the President, by and with the advice and consent of the Senate.

(b) The compensation of the Deputy Director shall be fixed by the President at a rate not in excess of the annual rate of compensation payable to the Deputy Director of the Bureau of the Budget.

(c) The Director is authorized to employ such officers and employees (including experts

* Pub. L. 91-224, 42 U.S.C. 4371-4374, April 3, 1970.

and consultants) as may be necessary to enable the Office to carry out its functions under this title and Public Law 91-190, except that he may employ no more than 10 specialists and other experts without regard to the provisions of title 5, United States Code, governing appointments in the competitive service, and pay such specialists and experts without regard to the provisions of chapter 51 and subchapter 111 of chapter 53 of such title relating to classification and General Schedule pay rates, but no such specialist or expert shall be paid at a rate in excess of the maximum rate of GS-18 of the General Schedule under section 5330 of title 5.

(d) In carrying out his functions the Director shall assist and advise the President on policies and programs of the Federal Government affecting environmental quality by —

(1) Providing the professional and administrative staff and support for the Council on Environmental Quality established by Public Law 91-190;

(2) Assisting the Federal agencies and departments in appraising the effectiveness of existing and proposed facilities, programs, policies, and activities of the Federal Government, and those specific major projects designated by the President which do not require individual project authorization by Congress, which affect environmental quality;

(3) Reviewing the adequacy of existing systems for monitoring and predicting environmental changes in order to achieve effective coverage and efficient use of research facilities and other resources;

(4) Promoting the advancement of scientific knowledge of the effects of actions and technology on the environment and encourage the development of the means to prevent or reduce adverse effects that endanger the health and well-being of man;

(5) Assisting in coordinating among the Federal departments and agencies those programs and activities which affect, protect, and improve environmental quality;

(6) Assisting the Federal departments and agencies in the development and interrelationship of environmental quality criteria and standards established through the Federal Government;

(7) Collecting, collating, analyzing, and interpreting data and information on environmental quality, ecological research, and evaluation.

(e) The Director is authorized to contract with public or private agencies, institutions, and organizations and with individuals without regard to sections 3618 and 3709 of the Revised Statutes (31 U.S.C. 529; 41 US.C. 5) in carrying out his functions.

Report

Sec. 204. Each Environmental Quality Report required by Public Law 91-190 shall, upon transmittal to Congress, be referred to each standing committee having jurisdiction over any part of the subject matter of the Report.

Authorization

Sec. 205. There are hereby authorized to be appropriated not to exceed $500,000 for the fiscal year ending June 30, 1970, not to exceed $750,000 for the fiscal year ending June 30, 1971, not to exceed $1,250,000 for the fiscal year ending June 30, 1972, and not to exceed $1,500,000 for the fiscal year ending June 30, 1973. These authorizations are in addition to those contained in Public Law 91-190.

Approved April 3, 1970.

THE CLEAN AIR ACT § 309*

§ 7609. Policy review

(a) The Administrator shall review and comment in writing on the environmental impact of any matter relating to duties and responsibilities granted pursuant to this chapter or other provisions of the authority of the Administrator, contained in any (1) legislation proposed by any Federal department or agency, (2) newly authorized Federal projects for construction and any major Federal agency action (other than a project for construction) to which section 4332(2)(C) of this title applies, and (3) proposed regulations published by any department or agency of the Federal Government. Such written comment shall be made public at the conclusion of any such review.

(b) In the event the Administrator determines that any such legislation, action, or regulation is unsatisfactory from the standpoint of public health or welfare or environmental quality, he shall publish his determination and the matter shall be referred to the Council on Environmental Quality.

Executive Order 11514, March 5, 1970

PROTECTION AND ENHANCEMENT OF ENVIRONMENTAL QUALITY

As amended by Executive Order 11991. (Sec. 2(g) and 3(h)). May 24, 1977**

By virtue of the authority vested in me as President of the United States and in furtherance of the purpose and policy of the National Environmental Policy Act of 1969 (Public Law No. 91-190, approved January 1, 1970), it is ordered as follows:

Section 1. *Policy*. The Federal Government shall provide leadership in protecting and enhancing the quality of the Nation's environment to sustain and enrich human life. Federal agencies shall initiate measures needed to direct their policies, plans and programs so as to meet national environmental goals. The Council on Environmental Quality, through the Chairman, shall advise and assist the President in leading this national effort.

Sec. 2. *Responsibilities of Federal agencies*. Consonant with Title I of the National Environmental Policy Act of 1969, hereafter referred to as the "Act", the heads of Federal agencies shall:

(a) Monitor, evaluate, and control on a continuing basis their agencies' activities so as to protect and enhance the quality of the environment. Such activities shall include those directed to controlling pollution and enhancing the environment and those designed to accomplish other program objectives which may affect the quality of the environment. Agencies shall develop programs and measures to protect and enhance environmental quality and shall assess progress in meeting the specific objectives of such activities. Heads of agencies shall consult with appropriate Federal, State and local agencies in carrying out their activities as they affect the quality of the environment.

* July 14, 1955, c.360, § 309, as added Dec. 31, 1970, Pub. L. 91-604 § 12(a), 42 U.S.C. § 7609 (1970).

** The Preamble to Executive Order 11991 is as follows:

By virtue of the authority vested in me by the Constitution and statutes of the United States of America, and as President of the United States of America, in furtherance of the purpose and policy of the National Environmental Policy Act of 1969, as amended (42 U.S.C. 4321 *et seq.*), the Environmental Quality Improvement Act of 1970 (42 U.S.C. 4371 *et seq.*), and Section 309 of the Clean Air Act, as amended (42 U.S.C. 1857h-7), it is hereby ordered as follows:

(b) Develop procedures to ensure the fullest practicable provision of timely public information and understanding of Federal plans and programs with environmental impact in order to obtain the views of interested parties. These procedures shall include, whenever appropriate, provision for public hearings, and shall provide the public with relevant information, including information on alternative courses of action. Federal agencies shall also encourage State and local agencies to adopt similar procedures for informing the public concerning their activities affecting the quality of the environment.

(c) Insure that information regarding existing or potential environmental problems and control methods developed as part of research, development, demonstration, test, or evaluation activities is made available to Federal agencies, States, counties, municipalities, institutions, and other entities, as appropriate.

(d) Review their agencies' statutory authority, administrative regulations, policies, and procedures, including those relating to loans, grants, contracts, leases, licenses, or permits, in order to identify any deficiencies or inconsistencies therein which prohibit or limit full compliance with the purposes and provisions of the Act. A report on this review and the corrective actions taken or planned, including such measures to be proposed by the President as may be necessary to bring their authority and policies into conformance with the intent, purposes, and procedures of the Act, shall be provided to the Council on Environmental Quality not later than September 1, 1970.

(e) Engage in exchange of data and research results, and cooperate with agencies of other governments to foster the purposes of the Act.

(f) Proceed, in coordination with other agencies, with actions required by section 102 of the Act.

(g) In carrying out their responsibilities under the Act, and this order, comply with the regulations issued by the Council except where such compliance would be inconsistent with statutory requirements.

Sec. 3. *Responsibilities of Council on Environmental Quality.* The Council on Environmental Quality shall:

(a) Evaluate existing and proposed policies and activities of the Federal Government directed to the control of pollution and the enhancement of the environment and to the accomplishment of other objectives which affect the quality of the environment. This shall include continuing review of procedures employed in the development and enforcement of Federal standards affecting environmental quality. Based upon such evaluations the Council shall, where appropriate, recommend to the President policies and programs to achieve more effective protection and enhancement of environmental quality and shall, where appropriate, seek resolution of a significant environmental issues.

(b) Recommend to the President and to the agencies priorities among programs designed for the control of pollution and for enhancement of the environment.

(c) Determine the need for new policies and programs for dealing with environmental problems not being adequately addressed.

(d) Conduct, as it determines to be appropriate, public hearings or conferences on issues of environmental significance.

(e) Promote the development and use of indices and monitoring systems (1) to assess environmental conditions and trends, (2) to predict the environmental impact of proposed public and private actions, and (3) to determine the effectiveness of programs for protecting and enhancing environmental quality.

(f) Coordinate Federal programs related to environmental quality.

(g) Advise and assist the President and the agencies in achieving international cooperation for dealing with environmental problems, under the foreign policy guidance of the Secretary of State.

(h) Issue regulations to Federal agencies for the implementation of the procedural provisions of the Act (42 U.S.C. 4332(2)). Such regulations shall be developed after consultation with affected agencies and after such public hearings as may be appropriate. They will be designed to make the environmental impact statement process more useful to decisionmakers and the public; and to reduce paperwork and the accumulation of extraneous background data, in order to emphasize the need to focus on real environmental issues and alternatives. They will require impact statements to be concise, clear, and to the point, and supported by evidence that agencies have made the necessary environmental analyses. The Council shall include in its regulations procedures (1) for the early preparation of environmental impact statements, and (2) for the referral to the Council of conflicts between agencies concerning the implementation of the National Environmental Policy Act of 1969, as amended, and Section 309 of the Clean Air Act, as amended, for the Council's recommendation as to their prompt resolution.

(i) Issue such other instructions to agencies, and request such reports and other information from them, as may be required to carry out the Council's responsibilities under the Act.

(j) Assist the President in preparing the annual Environmental Quality Report provided for in section 201 of the Act.

(k) Foster investigations, studies, surveys, research, and analyses relating to (i) ecological systems and environmental quality, (ii) the impact of new and changing technologies thereon, and (iii) means of preventing or reducing adverse effects from such technologies.

Sec. 4. *Amendments of E.O. 11472*. Executive Order No. 11472 of May 29, 1969, including the heading thereof, is hereby amended:

(1) By substituting for the term "the Environmental Quality Council", wherever it occurs, the following: "the Cabinet Committee on the Environment".

(2) By substituting for the term "the Council", wherever it occurs, the following: "the Cabinet Committee".

(3) By inserting in subsection (f) of section 101, after "Budget,", the following: "the Director of the Office of Science and Technology,".

(4) By substituting for subsection (g) of section 101 the following:

"(g) The Chairman of the Council on Environmental Quality (established by Public Law 91-190) shall assist the President in directing the affairs of the Cabinet Committee."

(5) By deleting subsection (c) of section 102.

(6) By substituting for "the Office of Science and Technology", in section 104, the following: "the Council on Environmental Quality (established by Public Law 91-190)".

(7) By substituting for "(hereinafter referred to as the 'Committee')", in section 201, the following: "(hereinafter refeıred to as the 'Citizens' Committee')".

(8) By substituting for the term "the Committee", wherever it occurs, the following: "the Citizens' Committee".

INDEX

A

B

C

D

E

O

P

R